CASTING TITANIUM ALLOYS

WITH SPECIAL REFERENCE TO AEROSPACE APPLICATIONS

Prepared By

P. A. Kobryn
Allan W. Gunderson
and Walter M. Griffith

Metals and Ceramics Division
Wright Laboratory

Revised by

Mark Watson

Wexford Press
2008

Table of Contents

List of Figures

List of Tables

1.   Introduction

Cast titanium alloys are practically unique in that they have identical, or nearly identical, strength compared to wrought titanium alloy parts. Because of the reduced cost in casting metal components, this makes casting a clear choice for mass production of any titanium alloy part.

The use of titanium and its alloys has grown tremendously over the past several decades. Properties such as superior corrosion resistance and high specific strength combined with progress in the areas of titanium production from ore, alloy development, and materials processing have led to the widespread use of titanium alloys within many industries. The aerospace industry remains the largest user of titanium, with both structural and turbine engine components commonly fabricated of titanium alloys. Other industries that frequently use titanium include the marine, power generation, chemical processing, biomedical, and sporting goods industries (most notably golf club heads.) Table 1 lists some applications of titanium and titanium alloy products.

Table 1.   Applications for titanium and titanium alloys

Application area	Typical uses
Aerospace	
Airframes	Fittings, bolts, landing gear beams, wing boxes, fuselage frames, flap tracks, slat tracks, brake assemblies, fuselage panels, engine support mountings, undercarriage components, inlet guide vanes, wing pivot lugs, keels, firewalls, fairings, hydraulic tubing, deicing ductings, SPF parts
Engines	Compressor disks and blades, fan disks and blades, casings, afterburner cowlings, flange rings, spacers, bolts, hydraulic tubing, hot-air ducts, helicopter rotor hubs
Satellites, rockets	Rocket engine casings, fuel tanks
Chemical processing	Storage tanks, agitators, pumps, columns, frames, screens, mixers, valves, pressurized reactors, filters, piping and tubing, heat exchangers, electrodes and anode baskets for metal and chlorine-alkali electrolysis
Energy Industry	
Power generating plants	Condensers, cooling systems, piping and tubing, turbine blades, generator retaining rings, rotor slot wedges, linings for FGD units, nuclear waste disposal
Geothermal energy	Heat exchangers, evaporators, condensers, tubes
Marine engineering	
Shipbuilding	Heat exchangers, condensers, piping and tubing, propellers, propeller and rudder shafts, data logging equipment, gyrocompasses, thruster pumps, lifeboat parts, radar components, cathodic protection anodes, hydrofoil struts
Diving equipment	Deep-sea pressure hulls, submarines (Soviet Union), submarine ball valves (United States)
Seawater desalination	Vapor heaters, condensers, thin-walled tubing

Application area	Typical uses
Offshore installations	Cooling equipment, condensers, heat exchangers, piping and tubing, flanges, deep-drilling riser pipes, flexible risers, desulfurizers, catalytic crackers, sour water strippers, regenerators, structural components
Biomedical engineering	Hip- and knee-joint prostheses, bone plates, screws and nails for fractures, pacemaker housings, heart valves, instruments, dentures, hearing aids, high-speed centrifugal separators for blood, wheelchairs, insulin pumps
Deep drilling	Drill pipes, riser pipes, production tubulars, casing liners, stress joints, instrument cases, wire, probes
Automotive industry	Connecting rods, valves, valve springs and retainers, crankshafts, camshafts, drive shafts, torsion bars, suspension assemblies, coil springs, clutch components, wheel hubs, exhaust systems, ball and socket joints, gears
Machine tools	Flexible tube connections, protective tubing, instrumentation and control equipment
Pulp and paper	Bleaching towers, pumps, piping and tubing
Food processing	Tanks (dairies, beverage industry), heat exchangers, components for packaging machinery
Construction	Facing and roofing, concrete reinforcement, monument refurbishing (Acropolis), anodes for cathodic protection
Superconductors	Wire rod of Ti-Nb alloys for manufacture of powerful electromagnets, rotors for superconductive generators
Fine art	Sculptures, fountain bases, ornaments, doorplates
Consumer products Jewelry industry	Jewelry, clocks, watches
Optics	Eyeglass frames, camera shutters
Sports equipment	Bicycle frames, tennis rackets, shafts and heads for golf clubs, mountain climbing equipment (ice screws, hooks), luges, bobsled components, horse shoes, fencing blades, target pistols
Musical instruments	Harmonica reeds, bells
Personal security and safety	Armor (cars, trucks, helicopters, fighter aircraft), helmets, bulletproof vests, protective gloves
Transportation	Driven wheelsets for high-speed trains, wheel tires
Cutting implements	Scissors, knives, pliers
Shape-memory alloys	Nickel-titanium alloys for springs and flanges
Miscellaneous	Pens, nameplates, telephone relay mechanisms, pollution-control equipment, titanium-lined vessels for salt-bath nitriding of steel products

1.1. Physical Metallurgy of Titanium Alloys

1.1.1. Alloying

Titanium alloys are often classified by their crystal structure. There are many different classes of commercial titanium alloys, including alpha, alpha-beta, beta, alpha-2, and gamma. Pure titanium has two allotropes- alpha, a low temperature hexagonal close packed phase, and beta, a high temperature body centered cubic phase. Low alloy content alloys typically consist of solid solution alpha and/or beta phase. Through the addition of different alloying elements, alpha, alpha-beta, beta, alpha-2, or gamma alloys can be formed. Alpha-2 and gamma alloys are based on the intermetallic phases Ti_3Al and $TiAl$, respectively. The properties of materials are strongly dependent on their crystal structure, so alloys of the same class tend to have similar properties.

Alloying elements are added for various purposes. Many alloying additions act to stabilize either the alpha or the beta phase. Alpha stabilizers favor the alpha crystal structure and increase the alpha + beta $\rightarrow$ beta transus temperature. Aluminum, gallium, germanium, carbon, oxygen, and nitrogen are all alpha stabilizers. Beta stabilizers favor the beta structure and decrease the beta transus temperature. Beta stabilizers can be beta isomorphous (molybdenum, vanadium, tantalum, and niobium) or beta eutectoid (manganese, iron, chromium, cobalt, nickel, copper, and silicon). Beta isomorphous elements dissolve in the beta phase, while beta eutectoid elements form [alpha + a compound] eutectoids with eutectoid temperatures significantly below the beta transus temperatures.

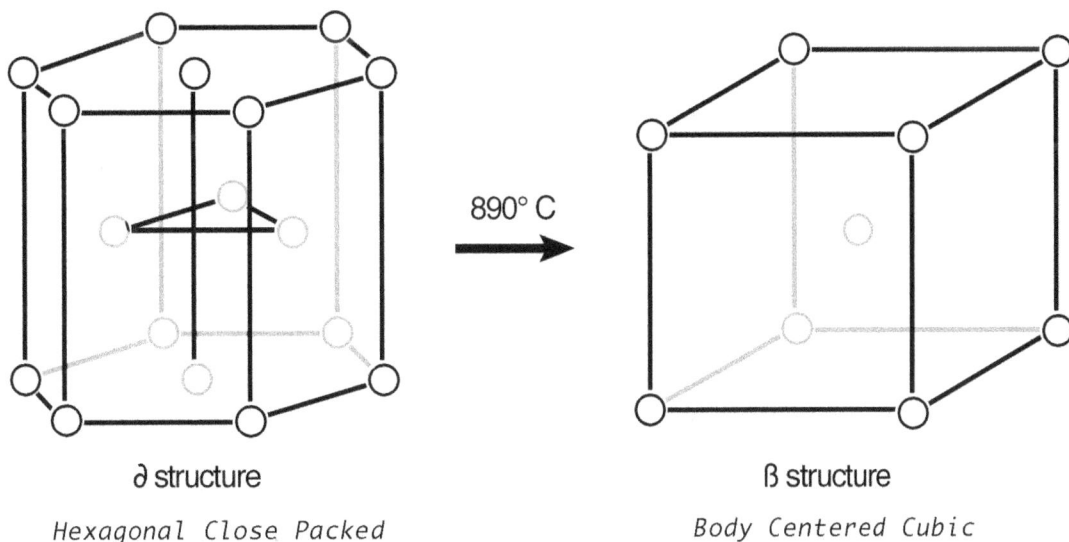

∂ structure

Hexagonal Close Packed

890° C →

ß structure

Body Centered Cubic

At around 890°C pure titanium undergoes an allotropic transformation from alpha to beta phase. Alloys may be either alpha, beta, or a mixture of the two phases.

Alloying additions can also be used to alter mechanical properties or corrosion resistance. Aluminum is an alpha stabilizer which also increases tensile and creep strength and elastic modulus. Strengthening in titanium-aluminum alloys is achieved through solid solution strengthening. Aluminum additions larger than 6% can also lead to the formation of Ti_3Al, which is a brittle intermetallic. Tin is another alpha stabilizer and solid solution strengthener, and can also form brittle intermetallic phases if present in large quantities. Zirconium is a weak beta stabilizer that retards transformation rates and increases strength by forming a continuous solid solution, but additions of more than 5 or 6% can reduce ductility and creep strength. Niobium is a beta stabilizer that improves high temperature oxidation resistance. Iron is a beta stabilizer that reduces creep strength. Carbon is an alpha stabilizer that increases the size of the alpha + beta phase field and increases tensile and fatigue strength. Small additions of palladium or molybdenum and nickel improve corrosion resistance.

The levels of residual impurities can also affect properties of titanium alloys such as lattice parameters, transformation temperatures, and mechanical properties. Residual carbon, nitrogen, silicon, iron, and oxygen raise strength but lower ductility. These elements are sometimes used as alloying elements in special high strength alloy grades, but are otherwise held to low levels. Special extra low interstitial (ELI) grades contain especially low amounts of interstitial impurities, and are used in applications requiring good ductility and toughness. Table 2[2] lists some titanium alloying additions and their effect on structure, while Figure 1[2] shows the main characteristics of several different alloy families.

Table 2.   Ranges and effects of some alloying elements used in titanium

Alloying Element	Range (approx.) wt %	Effect on structure
Aluminum	2 - 7	α stabilizer
Tin	2 - 6	α stabilizer
Vanadium	2 - 20	β stabilizer
Molybdenum	2 - 20	β stabilizer
Chromium	2 - 12	β stabilizer
Copper	2 - 6	β stabilizer
		α and β strengthener
Zirconium	2 - 8	solid solution strengthener
Silicon	0.05 - 1	Improves creep resistance

4

Ti 834 Ti-6Al-2Sn-4Zr-2Mo-0.8Si Ti 17

TA5E IMI 685 Ti-6Al-4V Ti-6Al-2Sn-4Zr-2Mo Betacez

β transus Forming capacity

Flow stress

Strain rate sensitivity

Weldability Heat treatment capacity

High temperature strength Room temperature strength

∂ alloys | Near-∂ alloys | ∂ + β alloys | Near β alloys | β alloys

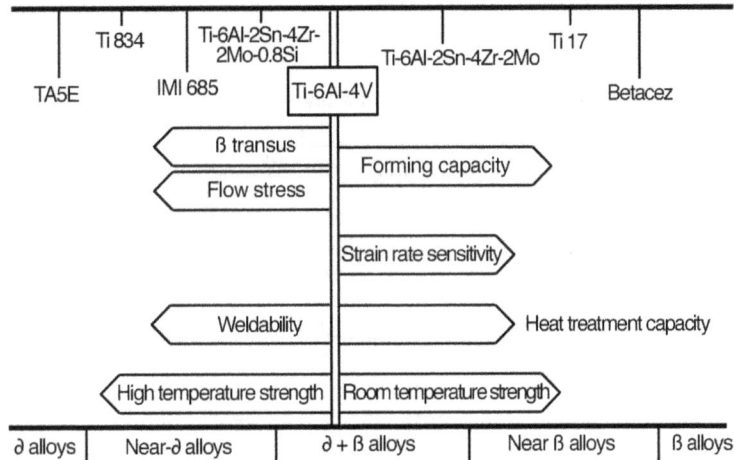

Figure 1. Main characteristics of the different titanium alloy families.

Commercially-pure (CP) titanium has an alpha crystal structure. CP Ti has good strength, but is weaker than titanium alloys or steel. Its modulus of elasticity is in the intermediate range, and is affected by texture. Its impact toughness is comparable to low alloy steel. Above 315°C, creep strength concerns begin. CP Ti has excellent corrosion resistance.

Alpha alloys have higher strength but less corrosion resistance than CP titanium. Alpha alloys have good ductility, even at low temperatures. Heat treatment is not a viable strengthening mechanism (because alpha is stable at room temperature), so grain size and cold work determine strength. Alpha alloys are not as forgeable as other titanium alloys. Most alpha alloys contain aluminum as the principal alloying element.

Alpha-beta alloys contain a combination of the alpha and beta phases, the distribution of which depends on composition and thermomechanical history. Because the phase morphology and distribution can greatly affect mechanical properties, alpha-beta alloys have properties that are strongly dependent on thermomechanical history. The effect of the processing temperature regime on the properties of alpha-beta alloys is shown qualitatively in Table 3.[2] Alpha-beta alloys are heat treatable, and can be strengthened by solution treatment and aging. Strength and ductility are inversely related, and can be adjusted by heat treatment. Some typical titanium heat treatments are listed in Table 4.[2] Ti-6Al-4V, an alpha-beta alloy, is the most common titanium alloy.

Table 3. Qualitative comparison of β processed and α+β processed titanium alloys

Property	β processed	α/β processed
Tensile strength	Moderate	Good
Creep strength	Good	Poor
Fatigue strength	Moderate	Good
Fracture toughness	Good	Poor
Crack growth rate	Good	Moderate
Grain size	Large	Small

Table 4. Summary of heat treatments for alpha-beta titanium alloys

Heat treatment designation	Heat treatment cycle	Microstructure
Duplex anneal	Solution treat at 50 - 75"C below T_β(a), air cool and age for 2 - 8 h at 540 - 675"C	Primary α, plus Widmanstätten α + β regions
Solution treat and age	Solution treat at ~40"C below T_β, water quench(b) and age for 2 - 8 h at 535 - 675"C	Primary α, plus tempered α' or an α + β mixture
Beta anneal	Solution treat at ~15"C above T_β, air cool and temper at 650 - 760"C for 2 h	Widmanstätten α + β colony microstructure
Beta quench	Solution treat at ~15"C above T_β,, water quench and stabilize at 650 - 760"C for 2 h	Tempered α'
Recrystallization anneal	925"C for 4 h, cool at 50"C/h to 760"C, air cool	Equiaxed α with β at grain-boundary triple points
Mill anneal	α + β hot work + anneal at 705"C for 30 min. to several hours and air cool	Incompletely recrystallized α with a small volume fraction of small β particles

(a) T_β is the β-transus for the particular alloy in question. (b) In more heavily β-stabilized alloys such as Ti-6Al-2Sn-4Zr-6Mo or Ti-6Al-6V-2Sn, solution treatment is followed by air cooling. Subsequent aging causes precipitation of a phase to form an α + β mixture.

Beta alloys have higher fracture toughness, better room temperature formability, higher yield strength, and better heat treatability than alpha-beta alloys. Beta alloys are often metastable, and can be hardened through the controlled precipitation of the stable alpha phase.

A special class of titanium alloys is the titanium aluminide class. Titanium aluminides are subdivided into alpha-2 and gamma titanium aluminides. Alpha-2 is an ordered Ti$_3$Al intermetallic phase, while gamma is an ordered TiAl intermetallic phase. These ordered intermetallics are often brittle at room temperature, but their superior high temperature properties make them attractive candidates for high temperature applications.

Table 5[2] lists some commercial titanium alloys with their structures, typical applications, and special properties.

Table 5.   Typical applications of various titanium-base materials

Nominal contents and common name or specification	Available mill forms	General description	Typical applications
Commercially pure titanium			
Unalloyed titanium	Bar, billet, extrusions, plate, strip, wire, rod, pipe, tubing, castings	For corrosion resistance in the chemical and marine industries, and where maximum ease of formability is desired. Weldability: good	Jet engine shrouds, cases, airframe skins, firewalls, and other hot-area equipment for aircraft and missiles; heat exchangers; corrosion resistant equipment for marine and chemical-processing industries. Other applications requiring good fabricability, weldability, and intermediate strength in service
Ti-0.2Pd: ASTM grades 7 and 11	Bar, billet, extrusions, plate, strip, wire, pipe, tubing, castings	The Pd-containing alloys extend the range of application in HCl, H$_3$PO$_4$, and H$_2$SO$_4$ solutions. Characteristics of good fabricability, weldability, and strength level are similar to those of corresponding unalloyed titanium grades.	For corrosion resistance in the chemical industry where media are mildly reducing or vary between oxidizing and reducing
Ti-0.3Mo-0.8Ni: ASTM grade 12	Bar, billet, extrusions, plate, strip, wire, pipe, tubing, castings	Compared to unalloyed Ti, Ti-0.3Mo-0.8Ni has better corrosion resistance and higher strength. The alloy is particularly resistant to crevice corrosion in hot brines.	For corrosion resistance in the chemical industry where media are mildly reducing or vary between oxidizing and reducing
α alloys			
Ti-2.5Cu: AECMA Ti-P11, or IMI 230	Bar, billet, rod, wire, plate, sheet, extrusions	Ti-2.5Cu combines the formability and weldability of titanium with improved mechanical properties from precipitation strengthening.	Useful for its improved mechanical properties, particularly up to 350°C (650°F). Aging doubles elevated-temperature properties and increases room-temperature strength by 25%.
Ti-5Al-2.5Sn (UNS R54520)	Bar, billet, extrusions, plate, sheet, wire, castings	Air frame and jet engine applications requiring good weldability, stability, and strength at elevated temperatures	Gas turbine engine casings and rings, aerospace structural members in hot spots, and chemical-processing equipment that require good weldability and intermediate strength at service temperatures up to 480°C (900°F)
Ti-5Al-2.5Sn-ELI (UNS R54521)	Same as UNS R54520	Reduced level of interstitial impurities improves ductility and toughness.	High-purity grade for pressure vessels for liquefied gases and other applications requiring better ductility and toughness, particularly in hardware for service to cryogenic temperatures

7

Nominal contents and common name or specification	Available mill forms	General description	Typical applications
Near-α alloys			
Ti-8Al-1Mo-1V (UNS R54810)	Bar, billet, extrusions, plate, sheet, wire, forgings	Near-α or α-β microstructure (depending on processing) with good combination of creep strength and fatigue strength when processed high in the α-β region (that is, near the β transus)	Fan blades are main use; forgings for jet engine components requiring good creep strength at elevated temperatures (compressor disks, plates, hubs). Other applications where light, high strength, highly weldable material with low density is required (cargo flooring)
Ti-6Al-2Sn-4Zr-2Mo (Ti-6242, or UNS 54620)	Bar, billet, sheet, strip, wire, forgings	Used for creep strength and elevated-temperature service. Fair weldability	Forgings and flat-rolled products used in gas turbine engine and air-frame applications where high strength and toughness, excellent creep resistance, and stability at temperatures up to 450°C (840°F) are required
Ti-6Al-2Sn-4Zr-2Mo-0.1Si (Ti-6242S)	Same as UNS 54620 but also castings	Silicon imparts additional creep resistance.	Same as UNS 54620 but maximum-use temperature up to about 520°C (970°F)
Ti-6Al-2Nb-1Ta-0.8Mo (UNS R56210)	Plate, sheet, strip, bar, wire, rod		Plate for naval shipbuilding applications, submersible hulls, pressure vessels, and other high-toughness applications
Ti-2.25Al-11Sn-5Zr-1Mo (Ti-679, UNS R54790)	Forgings, bar, billet, plate		Jet engine blades and wheels, large bulkhead forgings, other applications requiring high-temperature creep strength plus stability and short-time strength
Ti-5Al-5Sn-2Zr-2Mo-0.25Si (Ti-5522S, UNS 54560)	Forged billet and bar, special products available in plate and sheet	Semicommercial; no longer used	Specified in MIL-T-9046 and MIL-T-9047
IMI-685 (Ti-6Al-5Zr-0.5Mo-0.2Si)	Rod, bar, billet, extrusions	Weldable medium-strength alloy	Alloy for elevated-temperature uses up to about 520°C (970°F)
IMI-829 (Ti-5.5Al-3.5Sn-3Zr-1Nb-0.3Mo-0.3Si)	Rod, bar, billet, extrusions	Weldable, medium-strength alloy with good thermal stability and high creep resistance up to 600°C (1110°F)	Alloy for elevated-temperature uses up to about 580°C (1075°F)
IMI-834 (Ti-5.8Al-4Sn-3.5Zr-0.7Nb-0.5Mo-0.3Si)		Weldable, high-temperature alloy with improved fatigue performance as compared to IMI 829 and 685	Maximum-use temperature up to about 590°C (1100°F)
Ti-1100		Elevated-temperature alloy	Maximum-use temperature of 590°C (1100°F)
α-β alloys			
Ti-6Al-4V (UNS R56400 and AECMA Ti-P63)	Bar, billet, rod, wire, plate, sheet, strip, extrusions	Ti-6Al-4V is the most widely used titanium alloy. It is processed to provide mill-annealed or β-annealed structures, and is sometimes solution treated and aged. Ti-6Al-4V has useful creep resistance up to 300°C (570°F) and excellent fatigue strength. Fair weldability	Ti-6Al-4V is used for aircraft gas turbine disks and blades. It is extensively used, in all mill product forms, for airframe structural components and other applications requiring strength at temperatures up to 315°C (600°F); also used for high-strength prosthetic implants and chemical-processing equipment. Heat treatment of fastener stock provides tensile strengths up to 1100 MPa (160 ksi).
Ti-6Al-4V-ELI (UNS R56401)	Same as UNS R56400	Reduced interstitial impurities improve ductility and toughness.	Cryogenic applications and fracture-critical aerospace applications.
Ti-6Al-7Nb (IMI-367)	Rod, bar, billet, extrusions	High-strength alloy with excellent biocompatibility	Surgical implant alloy

Nominal contents and common name or specification	Available mill forms	General description	Typical applications
Corona 5 (Ti-4.5Al-5Mo-1.5Cr)	Alloy researched for plate, forging, and superplastic forming sheet	Improved fracture toughness over Ti-6Al-4V with less restricted chemistry. Easier to work than Ti-6Al-4V	Once investigated as a possible replacement for Ti-6Al-4V in aircraft, but no longer considered of interest
Ti-6Al-6V-2Sn (UNS R56620)	Bar, billet, extrusions, plate, sheet, wire	In the forms of sheet, light-gage plate, extrusions, and small forgings, this alloy is used for airframe structures where strength higher than that of Ti-6Al-4V is required. Usage is generally limited to secondary structures, because attractiveness of higher strength efficiency is minimized by lower fracture toughness and fatigue properties.	Applications requiring high strength at temperatures up to 315°C (600°F). Rocket engine case airframe applications including forgings, fasteners. Limited weldability. Susceptible to embrittlement above 315°C (600°F)
Ti-8Mn (UNS R56080)	Sheet, strip, plate	Limited usage	Aircraft sheet and structural parts
Ti-7Al-4Mo (UNS R56740)	Bar and forgings	Limited usage	Jet engine disks, compressor blades and spacers, sonic horns
Ti-6Al-2Sn-4Zr-2Cr-2Mo-0.25Si	Sheet, plate, and bar or billet for forging stock	Should be considered for long-time load-carrying applications at temperatures up to 400°C (750°F) and short-time load-carrying applications. Limited weldability	Forgings in intermediate temperature range sections of gas turbine engines, particularly in disk and fan blade components of compressors
Ti-6Al-2Sn-4Zr-2Cr-2Mo (UNS R56260)	Forgings, sheet	Heavy section forgings requiring high strength, fracture toughness, and high modulus	Forgings and sheet for airframes
Ti-3Al-2.5V (UNS R56320)	Bar, tubing, strip	Normally used in the cold-worked stress-relieved condition	Seamless tubing for aircraft hydraulic and ducting applications; weldable sheet; mechanical fasteners
IMI 550 and 551	Rod, bar, billet, extrusions	High-strength alloys; IMI 551 has increased room-temperature strength due to higher tin contents than IMI 550.	Two high-strength alloys with useful creep resistance up to 400°C (750°F)
β alloys			
Ti-13V-11Cr-3Al (UNS R58010)	Sheet, strip, plate, forgings, wire	High-strength alloy with good weldability	High-strength airframe components and missile applications such as solid rocket motor cases where extremely high strengths are required for short periods of time. Springs for airframe applications. Very little use anymore
Ti-8Mo-8V-2Fe-3Al (UNS R58820)	Rod, wire, sheet, strip, forgings	Limited weldability	Rod and wire for fastening applications; sheet, strip, and forgings for aerospace structures
Ti-3Al-8V-6Cr-4Zr-4.5Sn (Beta C)	Sheet, plate, bar, billet, wire, pipe, extrusions, castings	High-strength alloy with excellent ductility not available in other β alloys. Excellent cold-working characteristics; fair weldability	Airframe high-strength fasteners, rivets, torsion bars, springs, pipe for oil industry and geothermal applications
Ti-11.5Mo-6Zr-4.5Sn (Beta III)	Not being produced anymore	Excellent forgeability and cold workability. Very good weldability	Aircraft fasteners (especially rivets) and sheet metal parts where cold formability and strength potential can be used to greatest advantage. Possible use in plate and forging applications where high-strength capability, deep hardenability, and resistance to stress corrosion are required and somewhat lower aged ductility can be accepted

Nominal contents and common name or specification	Available mill forms	General description	Typical applications
Ti-10V-2Fe-3Al	Sheet, plate, bar, billet, wire, forgings	The combination of high strength and high toughness available is superior to any other commercial titanium alloy. For applications requiring uniformity of tensile properties at surface and center locations	High-strength airframe components. Applications up to 315°C (600°F) where medium to high strength and high toughness are required in bar, plate, or forged sections up to 125 mm (5 in.) thick. Used primarily for forgings
Ti-15V-3Al-3Cr-3Sn (Ti-15-3)	Sheet, strip, plate	Cold formable β alloy designed to reduce processing and fabrication costs. Heat treatable to a tensile strength of 1310 MPa (190 ksi)	High-strength aircraft and aerospace components
Ti-5Al-2Sn-2Zr-4Mo-4Cr (Ti-17)	Forgings	α-rich near-β alloy that is sometimes classified as an α-β alloy. Unlike other β or near-β alloys, Ti-17 offers good creep strength up to 430°C (800°F).	Forgings for turbine engine components where deep hardenability, strength, toughness, and fatigue are important. Useful in sections up to 150 mm (6 in.)
Transage alloys	Sheet, plate, bar, forgings	Developmental	High-strength (Transage 134) and high-strength elevated-temperature (Transage 175) alloys

1.1.2. Microstructure

The prior beta grain size and morphology, the morphology of the alpha phase, and the amount of transformation products present are the main microstructural features of titanium alloys. These features strongly influence mechanical properties. The prior beta grain size and morphology of titanium alloys is dependent on processing conditions such as cooling rate, amount of cold work, working temperature, and heat treatment conditions. The alpha phase can have several different morphologies, depending mainly on the conditions of formation. Cold working can cause grain elongation. Equiaxed alpha grains are obtained by annealing cold or hot worked alloys above the recrystallization temperature. Primary alpha is alpha that was stable at elevated temperatures. The morphology of this alpha is different from alpha formed by a transformation from beta. The beta-to-alpha transformation is strongly dependent on conditions during cooling from the high temperature regime, including cooling rate and initial temperature.

Upon cooling from above the beta transus temperature, alpha begins to form on the beta grain boundaries, and can also nucleate and grow along one or several sets of preferred crystallographic planes in the beta. The structure that results can be acicular (aligned alpha plates) or Widmanstätten alpha (basket weave structure), plate-like alpha (wide, long grains) or serrated alpha (irregular grains with jagged boundaries). Alternatively, the transformation from beta to alpha can be martensitic, i.e., produced by a diffusionless transformation mechanism. Two different types of martensite can be formed this way. Alpha prime, or hexagonal martensite, is the most common, but alpha double prime, or orthorhombic martensite, can form in certain alloys. The appearance of these structures is similar to acicular alpha, but is more well-defined and has straight sides. Acicular alpha is the most commonly observed morphology of transformed beta. The effect of different morphologies on properties is shown in Table 6[2]. An illustration of the formation of a Widmanstätten structure is shown in Figure 2[3], and examples of different morphologies of the popular Ti-6Al-4V alloy are shown in Figures 3[3] and 4.[3]

Table 6. **Relative advantages of equiaxed and acicular morphologies in near-alpha and alpha-beta alloys**

Equiaxed:
Higher ductility and formability Higher threshold stress for hot-salt stress corrosion Higher strength (for equivalent heat treatment) Better low-cycle fatigue (initiation) properties
Acicular:
Superior creep properties Higher fracture-toughness values Slight drop in strength (for equivalent heat treatment) Superior stress-corrosion resistance Lower crack-propagation rates

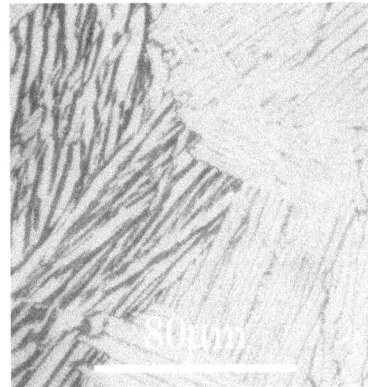

Photo: Bill Clyne

Figure 2. Formation of a Widmanstätten structure in a Ti-6Al-4V alloy. Achieved by cooling from above the ß transus. Final microstructure consists of plates of ∂ (white) separated by the ß phase (dark).

11

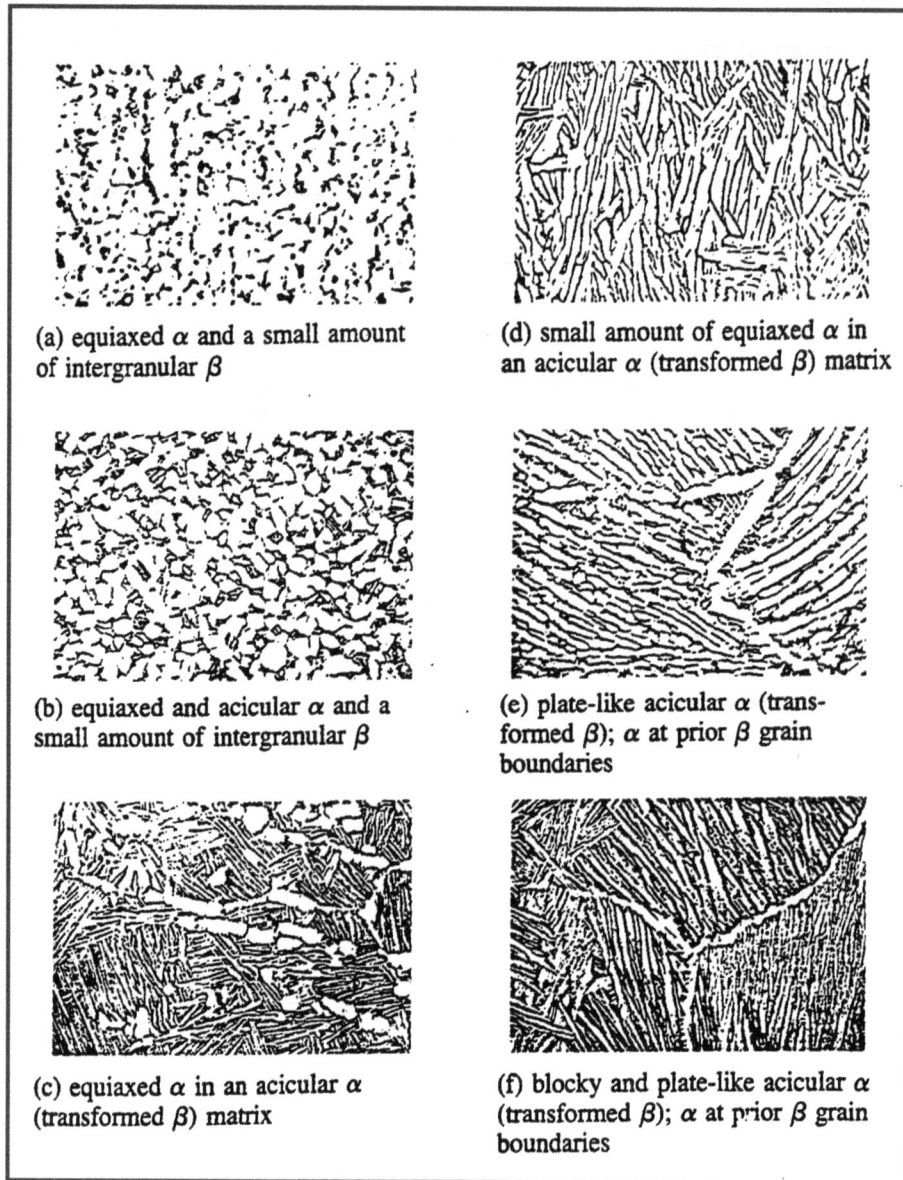

(a) equiaxed α and a small amount of intergranular β

(d) small amount of equiaxed α in an acicular α (transformed β) matrix

(b) equiaxed and acicular α and a small amount of intergranular β

(e) plate-like acicular α (transformed β); α at prior β grain boundaries

(c) equiaxed α in an acicular α (transformed β) matrix

(f) blocky and plate-like acicular α (transformed β); α at prior β grain boundaries

Figure 3. Optical microstructures of Ti-6Al-4V in six representative metallurgical conditions

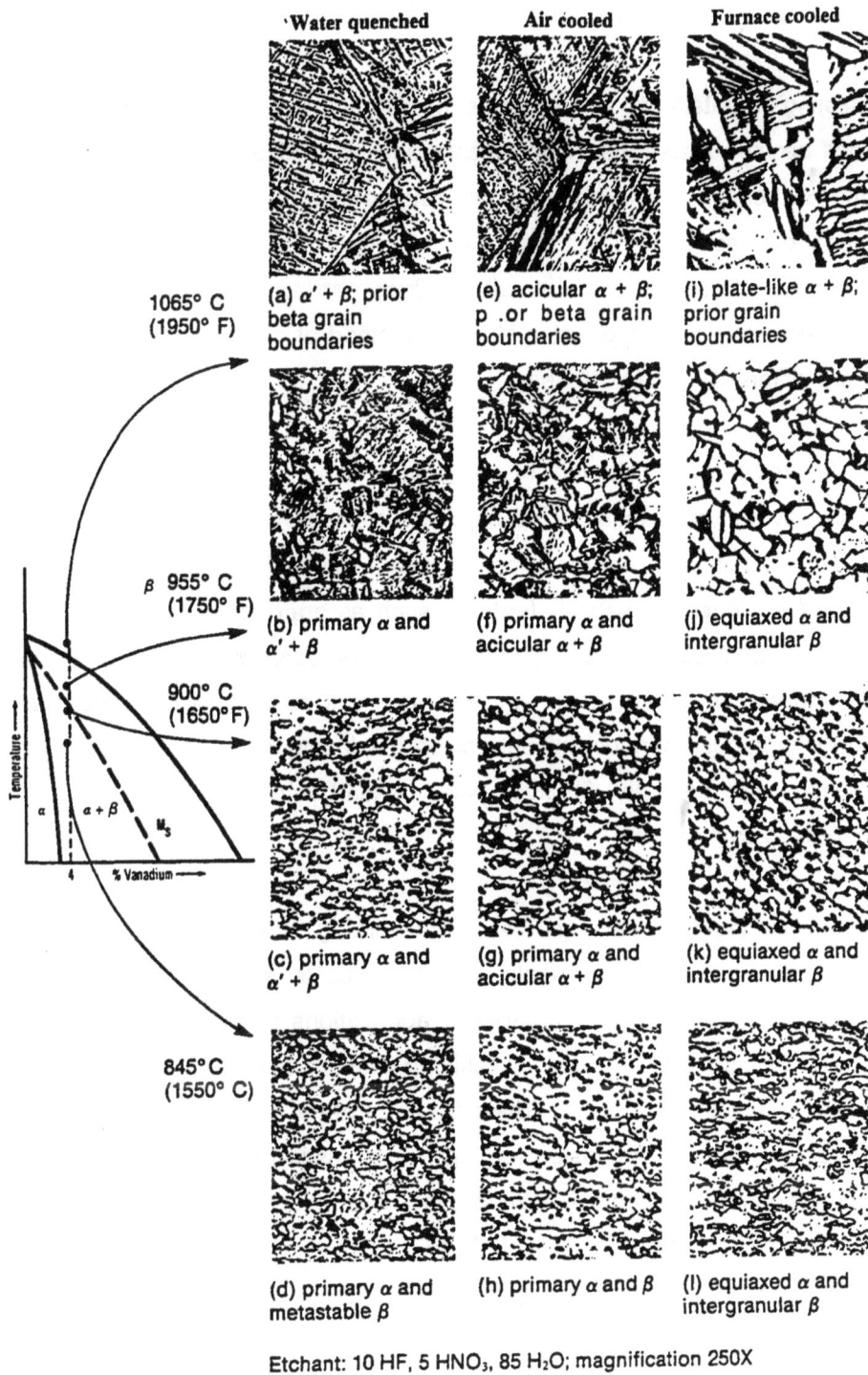

	Water quenched	Air cooled	Furnace cooled

1065° C
(1950° F)

955° C
(1750° F)

900° C
(1650° F)

845° C
(1550° C)

(a) $\alpha' + \beta$; prior beta grain boundaries

(e) acicular $\alpha + \beta$; p .or beta grain boundaries

(i) plate-like $\alpha + \beta$; prior grain boundaries

(b) primary α and $\alpha' + \beta$

(f) primary α and acicular $\alpha + \beta$

(j) equiaxed α and intergranular β

(c) primary α and $\alpha' + \beta$

(g) primary α and acicular $\alpha + \beta$

(k) equiaxed α and intergranular β

(d) primary α and metastable β

(h) primary α and β

(l) equiaxed α and intergranular β

Etchant: 10 HF, 5 HNO₃, 85 H₂O; magnification 250X

Figure 4. Ti-6Al-4V formation process on cooling.

13

1.2. Benefits of Casting vs. Wrought Processing

Casting of metals and alloys can be traced back to ancient times. Over the years, casting has become a well developed technology with diverse capabilities. There are many different casting methods, such as sand casting, investment casting, permanent mold casting, die casting, and lost foam casting. All of these casting processes involve the filling of a mold with molten metal or alloy. Upon cooling, the mold imparts shape to the solidifying material. The type of mold, the method of filling, and the method of part removal are different in each process.

Compared to other forming methods such as wrought processing or machining, casting has several advantages. The biggest advantage of casting over other forming methods is its flexibility. A large range of part sizes and complexities can be produced by casting. Also, microstructural features such as grain size, phase morphology, and porosity can be controlled. Casting also provides an economic advantage in many cases. Large assemblies can be reduced to single integral castings. Parts can be cast to near net shape and with special surface finishes, greatly reducing or eliminating final machining costs. Casting also typically involves shorter lead times from design to production, which decreases cost as well.

The advantages of casting over metalworking methods can be applied to the manufacture of titanium components. Forming costs and, more importantly, final machining costs greatly limit the number of applications for which the use of titanium is feasible.   The high cost of final machining makes net shape processing a viable alternative to conventional forming. The advantage of net shape techniques can be readily seen. Processes such as precision forging, superplastic forging, powder metallurgy, and casting have potential to reduce both forming and finishing costs for titanium alloys. Casting is the most well developed of these near net shape forming methods, and is commonly used to form titanium components.

Historically, reservations about the structural integrity of castings has limited their use for highly stressed components. Castings are often perceived as having inferior mechanical properties because of their inherent porosity and segregation. With the development of hot isostatic pressing (HIP) for closing internal porosity, cast + HIP'd + heat treated titanium castings having properties that meet or exceed those of forged components are being made (Table 7[4] and Figures 5[2], 6[4], and 7[4]).

14

The major limitation on titanium casting is related to its high melting temperature and reactivity in the molten state. Research into new melting and casting processes continues, with the hope that casting will significantly reduce the cost of manufacturing titanium components, thereby widening the range of applicability of titanium.

Table 7. Tensile properties and fracture toughness of Ti-6Al-4V cast coupons compared to typical wrought β-annealed material

Material conditions(a)	Yield strength		Ultimate tensile strength		Elongation, %	Reduction of area, %	K_{IC}	
	MPa	ksi	MPa	ksi			MPa√m	ksi√in
As-cast	896	130	1000	145	8	16	107	97
Cast HIP	869	126	958	139	10	18	109	99
BUS(b)	938	136	1041	151	8	12	---	---
GTEC(b)	938	136	1027	149	8	11	---	---
BST(b)	931	135	1055	153	9	15	---	---
ABST(b)	931	135	1020	148	8	12	---	---
TCT(b)	1055	153	1124	163	6	9	---	---
CST(b)	986	143	1055	153	8	15	---	---
KTH(b)	1055	153	1103	160	8	15	---	---
Typical wrought β annealed	860	125	955	139	9	21	91	83

(a) All conditions (except as-cast) are cast + HIP. (b) See reference for process details.

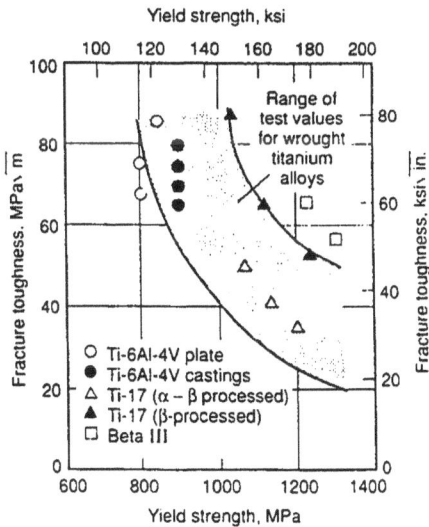

Figure 5. Fracture toughness of Ti-6Al-4V castings compared to Ti-6Al-4V plate and to other Ti alloys.

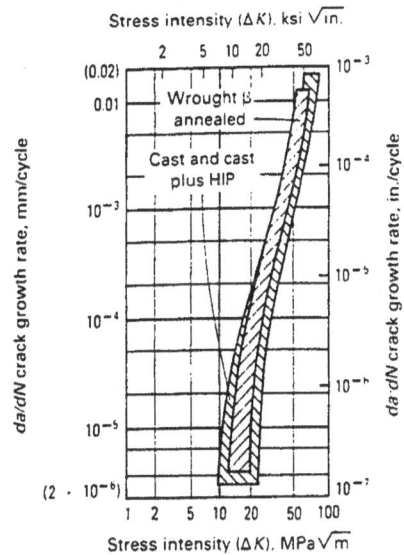

Figure 6. Scatterband comparison of FCGR behavior of wrought I/M β-annealed Ti-6Al-4V to cast and cast HIP Ti-6Al-4V data.

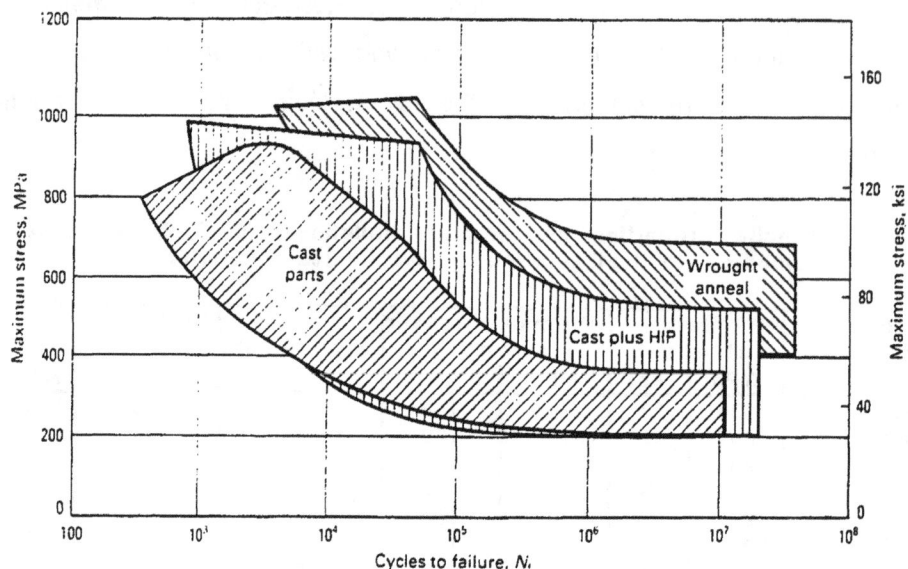

Figure 7. Comparison of smooth axial room-temperature fatigue rate in cast and wrought Ti-6Al-4V with R = +0.1.

1.3. A Brief Overview of This Report

This report contains four chapters. The present chapter introduces the topic of casting of titanium alloys and provides background information on the physical metallurgy of titanium, the benefits of casting, and a brief summary of the topics to be addressed in future chapters.

Chapter II reviews casting techniques for titanium alloys. First, the investment casting technique is discussed in detail, along with typical problems, defects, and limitations. Then permanent mold casting (PMC) is discussed. The process is described in detail, followed by a discussion of mold wear mechanisms and observed mold wear problems in PMC. The chapter concludes with a discussion of future research issues and approaches for titanium casting.

Chapter III addresses the concept of modeling of microstructural evolution during casting. First, background information on solidification (G versus R) maps is given; this summary includes how such maps are determined and how they are used. This background is followed by some examples of solidification maps for non-titanium systems. Some analytical microstructure modeling approaches are presented next,

16

followed by microstructure observations and interpretations in cast titanium alloys and rapidly solidified titanium powders. The chapter concludes with a discussion of future research issues and approaches to such research in the area of modeling of microstructural evolution.

Chapter IV outlines the modeling of solidification processes. First analytical modeling approaches are presented, followed by numerical methods and a review of some existing modeling software. Some examples of the use of solidification modeling to predict mold fill, shrinkage, and microstructure are presented next. Then the required input data for solidification modeling and the methodology for determination of such data are presented. Finally, future research issues and approaches in the area of modeling are discussed.

2. Casting of Titanium Alloys

Casting is currently used to produce many titanium alloy components. Casting of titanium alloys is a complicated endeavor because of the reactivity of titanium in the molten state. Hence, the application of casting to the manufacture of titanium parts has been paced to a great extent by the development of techniques to melt and contain the alloy without significant contamination. With the development of special melting techniques such as vacuum arc remelting (VAR) and induction skull remelting (ISR), casting of titanium has become feasible.

Casting of titanium alloys is done under vacuum to avoid oxygen and nitrogen pickup, and special mold materials are used to minimize reaction with the molten metal. Investment casting is the most popular casting method for titanium, although other methods are also used. These methods include rammed graphite, machined graphite, and more recently, permanent mold casting, although none of these techniques is used as frequently as investment casting. Figure 8[5] depicts a typical titanium casting process.

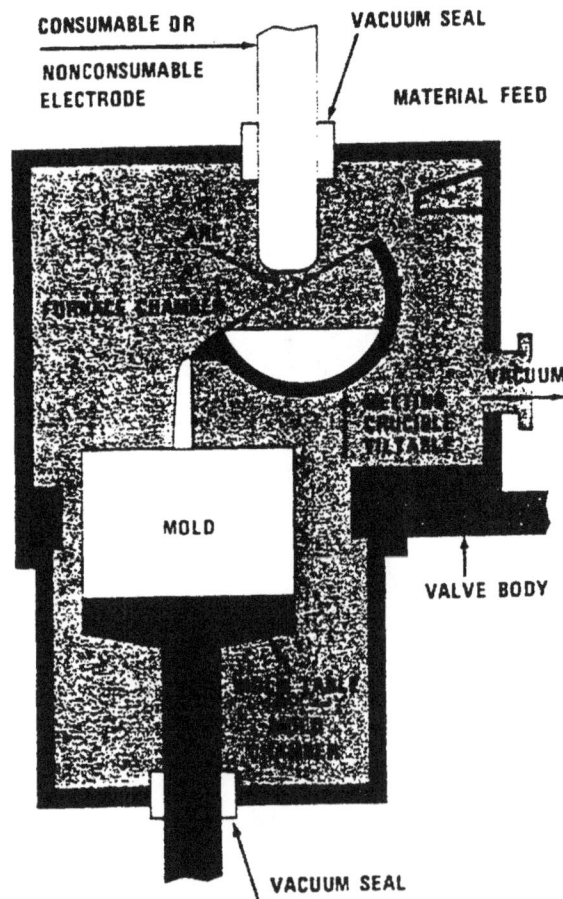

Figure 8. Schematic diagram of a titanium casting furnace.

This chapter describes the titanium casting procedure in detail. A review of the investment casting process as it applies to titanium is presented, with special attention given to the melting process, per se, and hot isostatic pressing (HIP) of castings. This is followed by a description of the permanent mold process, with special attention given to mold wear issues. Lastly, a discussion of future research issues and approaches in titanium casting is given.

2.1. Investment Casting

Investment casting is a method used to produce intricate components with fine features. It involves the use of a disposable ceramic mold produced by coating a wax pattern with successive layers of ceramic slurries and stuccos. There are many steps to the investment casting process, such as pattern making, mold making, pouring, and finishing. Figure 9[6] is a diagram of the steps in the investment casting process. Reference 7 gives an in-depth description of the investment casting process which is summarized here with special attention to titanium-specific issues.

2.1.1. Pattern Making

The first step to investment casting is pattern making. The patterns are made by injecting liquid, semi-solid or solid wax or plastic into a permanent mold. The compositions of the materials used for pattern making include wax blends (combinations of waxes, resins, plastics, fillers, and dyes) and plastics (polystyrene, polyethylene, etc.). In choosing a pattern material, properties affecting injection, removal, handling, assembly, dimensional control, mold making, dewaxing and burnout, economics, and environmental contamination must be considered. Wax blends are more commonly used than plastics because waxes are easy to process and can be formulated to produce a wide range of properties. Plastic patterns involve very low material costs, can be made on high production rate automatic injection machines, are easy to handle without damage, and have nearly infinite shelf lives, but are not as popular as wax patterns because of high tooling costs and the tendency to cause mold cracking. Plastics are most commonly used for small, thin, delicate parts that take advantage of the improved handling ability provided by plastics.

Reusable split molds are used to form investment casting patterns. The materials used for pattern molds include various steels, non-ferrous alloys, plastics, rubbers, and plasters. The choice of mold material depends on the injection system, the pattern material used, and the size and configuration of the pattern. Plastic patterns usually require the use of steel or beryllium-copper molds. Soft alloy and non-metallic molds are used for small production runs or for temporary tooling. Rubber molds are commonly used for jewelry patterns.

INVESTMENT CAST PROCESS
Sequence of Operations

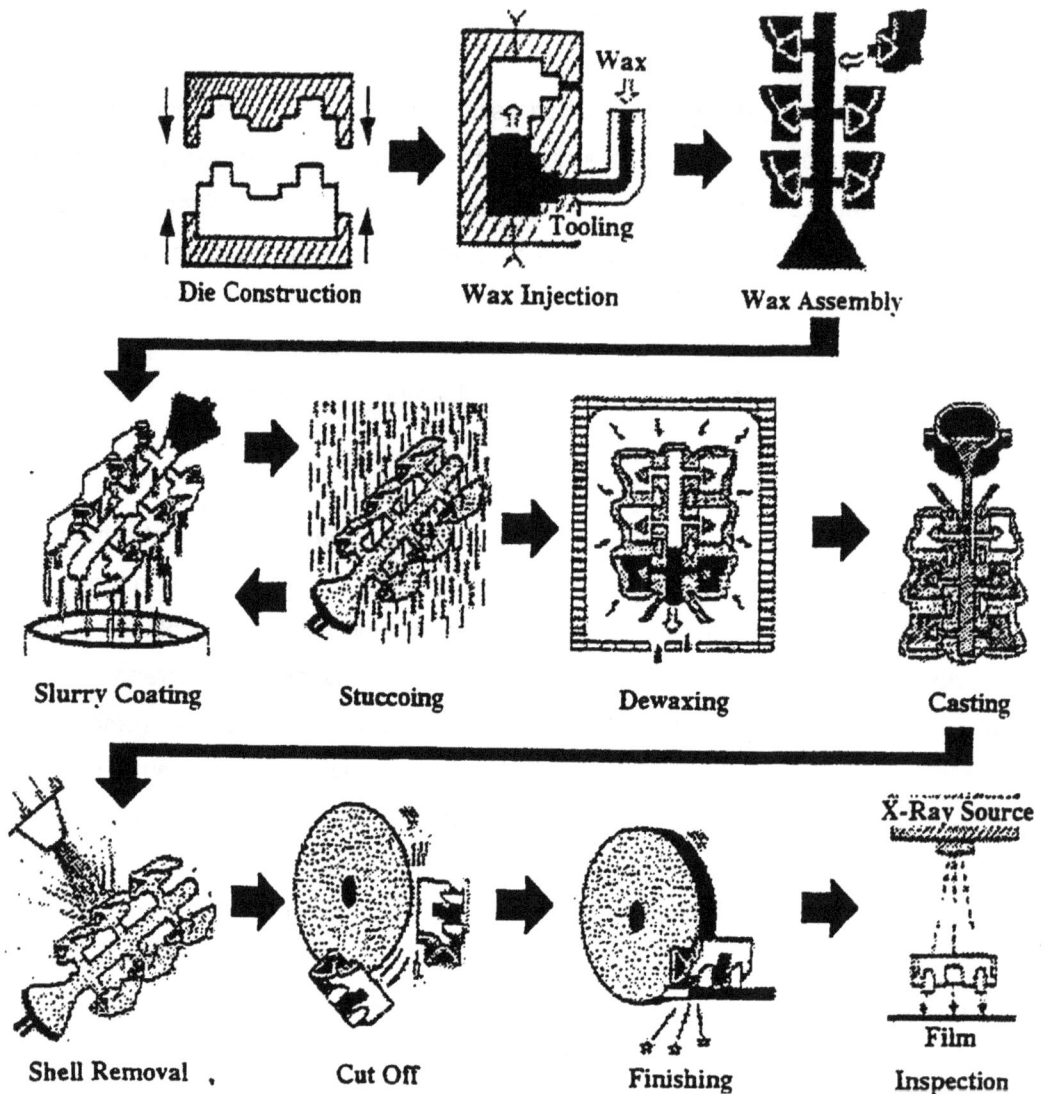

Figure 9. Conventional investment casting process showing primary processing steps of mold manufacture, casting, post-cast processing and inspection.

20

The injection system can be hand-operated or automated, pneumatic or hydraulic, and horizontal or vertical. Small shops can use small, simple, manual systems with little capital investment, while large shops requiring huge production capabilities can invest in large, automatic, hydraulic injection systems. Control of the injection process has a great effect on the resulting pattern, and so temperature, pressure, flow rate, and dwell time are usually monitored and controlled.

Once the patterns are produced, they must be prepared for the molding process. Larger patterns are sometimes cast individually, but smaller patterns are connected together in large clusters prior to mold making to minimize the number of processing steps required. Some extremely large or complex patterns are produced in separate pieces and must be connected before molding. Patterns produced without sprues, gates, and runners also require some pre-molding preparation.

The forming of a cluster of patterns is done using separately produced segments of pattern material. The configuration of the cluster is important from a mold making, mold filling, and solidification standpoint, and must be carefully designed to produce castings of the highest quality and lowest cost. The cluster must be strong enough to withstand the mold making process and must produce a mold of sufficient strength. Total runner volume within the cluster must be minimized to maximize casting yield. The cluster must also be configured to promote solidification toward the gating (to minimize shrinkage).

Pattern assembly is done with the use of wax welding for wax patterns and solvent welding for plastic patterns. Wax welding can be done using a hot iron, a spatula and a torch, or a laser welding system. The wax is melted at the surface of the parts to be attached, the two parts are pressed together, and the wax is allowed to resolidify. Solvent welding is done by applying solvent to the surfaces to be joined. The two softened surfaces are held in contact until hardening occurs. Most pattern assembly is done by hand and requires a fair amount of skill to ensure strong joints free of undercuts and to prevent damage to the patterns by dripping hot wax or solvent.

After the cluster is assembled, it is cleaned before mold making begins. Cleaning is done in (a) a solution containing a wetting agent, a non-attacking solvent, or a solvent that promotes adhesion through fine, uniform etching, (b) a liquid that deposits an extremely thin layer of refractory oxide to make the surface hydrophilic, or (c) a combination of these. Rinses are sometimes used to remove the cleanser, and the clusters are dried before the mold making process starts.

2.1.2. Mold Manufacture

To create the ceramic shell mold, the pattern or cluster is first dipped into a ceramic slurry bath containing different amounts of fine refractory powders, binders, wetting agents, and antifoam compounds. The pattern is removed from the slurry, allowed to drain momentarily, and then stuccoed with coarse ceramic powder in a fluidized bed or by sprinkling. The coating is hardened by drying or chemical gelling, and another slurry-stucco coating is applied. The process of dipping, stuccoing, and hardening is repeated until the desired mold thickness is reached. A final layer of slurry coating is applied without stuccoing to seal the mold. The slurry coats provide strength and duplicate the surface of the pattern exactly, while the stucco coats prevent the slurry from running off, cracking, or pulling away, provide bonding between adjacent layers, and build up mold thickness rapidly.

The specific type of refractory and binder used to make the mold depends on the alloy being cast and the process parameters being used (time at temperature, thermal gradients within the mold, etc.). The most common refractories used in investment casting are fused silica, zircon, and aluminum silicates, although alumina is frequently used for casting of superalloys because it is more refractory and less reactive. In titanium investment casting, zirconia, yttria, and thoria are commonly used.[6] Binders include colloidal silica, ethyl silicate, and liquid sodium silicate, as well as colloidal alumina and colloidal zirconia for processes requiring highly refractory binders such as directional solidification.

For the initial face coat, the refractory powder in the slurry is very fine and the viscosity of the slurry is high. The first stucco layer is applied with a fine refractory powder. The fine powder and high viscosity of the slurry produce a smooth, impenetrable surface. Wetting agents and antifoam compounds are sometimes added to the initial slurry to promote exact reproduction of the pattern surface without voids, and nucleating agents are added to refine the grain structure of the casting. Backup coats, in which porosity and surface finish are not important, are applied with coarser, less expensive powders and less viscous slurries.

After the final coat is applied, the shell is dried for 16 to 48 hours. Vacuum or low humidity atmospheres can enhance this final drying. Once the shell is completely dried, the pattern must be removed from the shell in a process termed dewaxing. There are two main types of dewaxing - autoclave and flash dewaxing. Both have the goal of rapidly melting the surface layer of wax before the rest of the pattern heats up. This allows the remaining wax to expand upon heating without putting undue stress on the mold, which expands at a slower rate.

In autoclave dewaxing, the wax is melted by rapid pressurization in less than 15 minutes, and the molten wax is collected for reuse. This method is the most popular dewaxing method, but is not applicable for polystyrene patterns. Flash dewaxing is done by placing the shell in a preheated furnace equipped with a drainage hole or an open bottom for the removal of the molten wax. As the wax drains from the furnace, some of it burns, but the majority is recovered for future use. This method can be used to burn out polystyrene patterns. A cheaper method of dewaxing is liquid dewaxing. This method does not require a large investment in equipment, but can represent a fire hazard.

Once the pattern is removed, the mold must be fired to remove excess moisture and to burn out organic binders, to impart strength by drying, and to preheat the mold for pouring to ensure good filling of fine details. Firing is usually done in a gas-fired furnace unless the casting technique being employed dictates the use of special equipment. Burnout, drying, and preheating can be conducted in one step or in separate steps. Burnout, drying, and preheat temperatures are dictated by the binders, refractories, and casting alloys used. For titanium casting, mold preheats range from 300 to 980°C.[6]

2.1.3. Casting

Once the mold is sintered and preheated, it is ready for the actual casting operation. Casting alloys such as steels, irons, cobalt alloys, and nickel alloys are usually melted with induction furnaces using magnesia, alumina, or zirconia crucibles and vacuum or air atmospheres. Gas fired and electrical resistance furnaces can be used for lower melting-point alloys such as aluminum, magnesium, and copper. Titanium alloys, due to their high melting point and reactivity, require more specialized melting techniques. All titanium melting techniques are done under vacuum or inert atmosphere to prevent oxygen and nitrogen pickup from air.

23

Vacuum arc remelting (VAR) is the most common method for melting titanium. In titanium VAR, an electric arc is used to melt a consumable titanium electrode into a water-cooled copper crucible. The first titanium to melt instantly solidifies on the copper crucible forming a protective skull. The skull remains solid throughout the melting and pouring, protecting the molten titanium from contamination. A schematic of the VAR process is shown in Figure 10.[8] Electron beam melting or induction melting can also be used to melt titanium, with the same type of cooled copper crucible to promote protective skull formation.

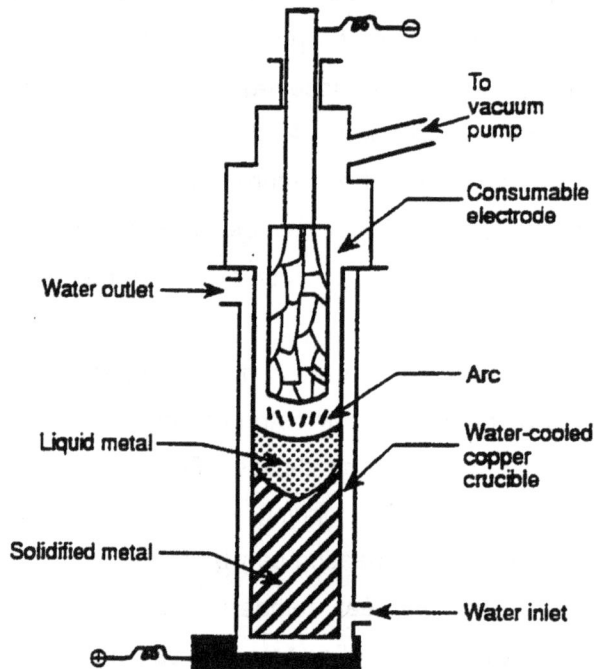

Figure 10. Schematic diagram of a vacuum-arc consumable-electrode furnace.

Once the molten charge is prepared and the mold is preheated, the casting process can begin. For conventional alloys, this can be done in air, but for reactive alloys such as titanium, vacuum or an inert atmosphere must be used. In both cases, the molten alloy is poured into the preheated mold and allowed to solidify. Most investment casting is done under the influence of gravity alone, although vacuum-assisted casting, centrifugal casting, and other pressurization techniques can also be used with investment casting.

2.1.4. Finishing

Once the casting is completely solidified and cooled, the ceramic mold is removed in a step called knockout. Most of the brittle mold is broken with a vibrating pneumatic hammer, abrasive or water blasting, or by hand. If the first refractory layer remains bonded to the casting after knockout, shotblasting or chemical means are used to remove it. Likewise, any cores that cannot be removed by mechanical means are removed by chemical dissolution.

Once the mold is removed, the gating must be removed from the individual components in a step called cut-off. Thin gates in brittle alloy castings are usually knocked off with a hammer. When use of a hammer is not feasible, band saws, cutting wheels, torches, or shear dies are used to remove gating. Grinding is used to smooth out the gates following cut-off. After gate removal, the castings are visually inspected for obvious defects. Any defective castings are discarded, and the remaining castings move on to finishing operations.

At this point, any required heat treatments are done. For example, titanium investment castings contain centerline shrinkage voids which are unacceptable in most applications. Consequently, a process known as hot isostatic pressing (HIP) is used to close internal porosity. HIP is carried out in a special high pressure vessel. For titanium alloys, the castings are placed inside the HIP vessel (Figure 11[9]), which is sealed, pressurized with ~100 MPa of argon, and heated to ~900 to 960°C.[6] The high pressure combined with the high temperature results in the elimination of internal voids. The pressure collapses the pores, which then heal by diffusion bonding.

At this point, any scale or harmful surface layers formed during the casting process or the HIP cycle must be removed. For example, reaction of the titanium with the mold during casting results in the formation of an oxygen enriched surface layer called an alpha case, which is removed via chemical milling in hydrofluoric acid solutions. Once brittle surface layers are removed, any surface defects must be repaired. For example, HIP generally results in the formation of surface depressions, which must be weld repaired. For titanium, repair is done in an argon atmosphere using gas tungsten arc welding with filler metal of the same alloy as the casting.

Once the casting is HIP'd, chemically milled, and repaired, any required finish machining and stress relieving or strengthening operations are performed. Finally, the casting is inspected. Inspection methods used include fluorescent penetrant inspection (FPI) to detect surface defects, radiography to detect internal porosity or density variations, hardness testing, and dimensional inspection.

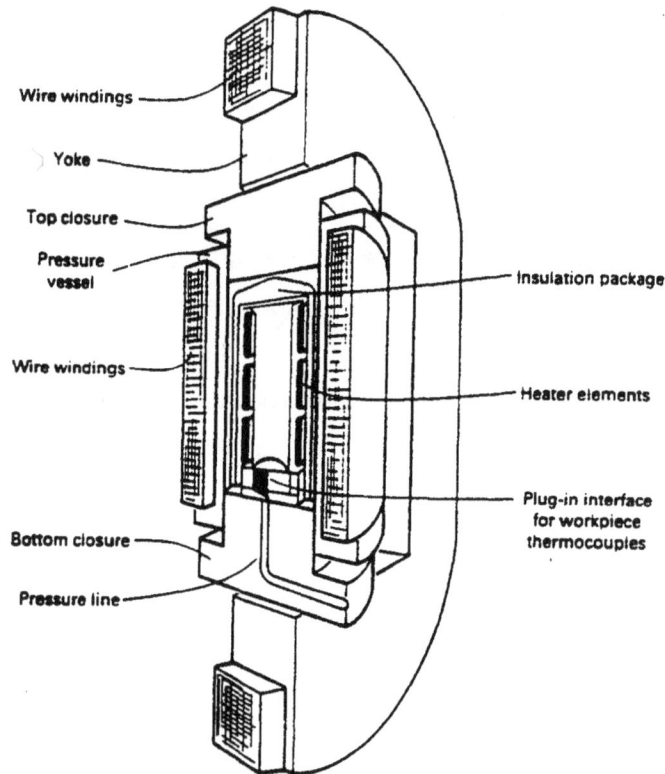

Figure 11. Schematic of one type of large, production HIP vessel.

2.1.5. Casting Defects

Common defects in titanium investment castings include centerline shrinkage voids, incomplete mold fill, inclusions, core and mold shift, cracking, and poor surface finish. Centerline shrinkage occurs when solidification does not proceed from one end of the casting to another, and can be repaired by HIP if it is not too severe. Incomplete mold fill occurs when the molten alloy is not fluid enough to completely fill the mold before freezing occurs. This problem can be corrected by adding more superheat to the metal, by increasing the mold preheat, or by using a centrifugal casting technique. Inclusions occur when the melt is "dirty" or when parts of the mold break off and become frozen within the casting. Inclusions can be eliminated by starting with a cleaner melt, by filtering the melt before mold filling, or by redesigning or increasing the strength of the mold. Core and mold shift cause dimensional variations and inaccuracies and can

26

occur either during mold making (from shifting of the wax patterns) or during mold filling (from the pressure of the molten alloy). Mold and core shift can be avoided by redesigning the mold to increase strength or by changing the fill pattern to avoid impingement on cores or unstable mold walls. Cracking occurs due to either mechanical stress imposed on the casting by the mold or thermal stress induced by thermal gradients within the casting. Cracking can be avoided by redesigning the mold to avoid constraining the casting or to change the temperature gradients within the casting. Poor surface finish occurs when the original surface finish of the mold is defective, when the molten alloy is very hot, or when the pour velocity is high enough to cause erosion of the original smooth mold surface. Poor surface finish can be avoided by reducing mold defects, eliminating unnecessary superheat, and controlling the flow pattern during filling.

2.1.6. Properties and Applications

The mechanical properties of titanium investment castings are usually similar to those of wrought titanium (Table 8[4]). Typical properties of investment cast + HIP'd Ti-6Al-4V include a UTS of 750 to 900 MPa, 0.2% YS of ~835 MPa, and an elongation to failure of 6 to 12%.[6] The fatigue strength of cast titanium, although greatly improved by HIP, does not compare well to the fatigue strength of wrought titanium. The poor fatigue properties of cast titanium are attributed to the large grain size and the presence of grain boundary alpha.

Table 8. Typical room temperature tensile properties of titanium alloy castings (bars machined from castings)
Specification minimums are less than these typical properties.

Alloys(a),(b)	Yield strength MPa	Yield strength ksi	Ultimate strength MPa	Ultimate strength ksi	Elongation, %	Reduction of area, %
Commercially pure (grade 2)	448	65	552	80	18	32
Ti-6Al-4V, annealed	855	124	930	135	12	20
Ti-6Al-4V ELI	758	110	827	120	13	22
Ti-1100, Beta-STA(c)	848	123	938	136	11	20
Ti-6Al-2Sn-4Zr-6Mo, annealed	910	132	1006	146	10	21
IMI-834, Beta-STA(c)	952	138	1069	155	5	8
Ti-6Al-2Sn-4Zr-6Mo, Beta-STA(c)	1269	184	1345	195	1	1
Ti-3Al-8V-6Cr-4Zr-4Mo, Beta-STA(c)	1241	180	1330	193	7	12
Ti-15V-3Al-3Cr-3Sn, Beta-STA(c)	1200	174	1275	185	6	12

(a) Solution-treated and aged (STA) heat treatments may be varied to produce alternate properties.
(b) ELI, extra low interstitial. (c) Beta-STA, solution treatment with β-phase field followed by aging.

Titanium castings can be used for structural components of all shapes and sizes, but parts with sections that are less than about 0.1" thick cannot be easily produced with investment casting.[6] Typical tolerances and surface finishes of titanium castings are given in Tables 9[4] & 10.[4] Titanium castings see limited use in cyclically loaded components in which fatigue strength is important. Also, titanium, whether wrought or cast, has an operating temperature limit of about 400°C. Titanium investment castings are commonly used in the aerospace industry. Gas turbine jet engine parts such as housings, casings, struts, and blades are commonly investment cast. Titanium castings are also used in marine applications, heat exchangers, medical prostheses, and other applications requiring superior corrosion resistance and strength. Some representative parts are shown in Figures 12 - 14.[4]

Table 9.   General linear and diametric tolerance guidelines for titanium castings

| Size | | Total tolerance band (a) | |
mm	in.	Investment cast	Rammed graphite process
25 to <102	1 to <4	0.76 mm (0.030 in.) or 1.0%, whichever is greater	1.52 mm (0.060 in.)
102 to <305	4 to <12	1.02 mm (0.040 in.) or 0.7%, whichever is greater	1.78 mm (0.70 in.) or 1.0%, whichever is greater
305 to <610	12 to <24	1.52 mm (0.060 in.) or 0.6%, whichever is greater	1.0%
≥610	≥24	0.5%	1.0%
Examples			
254 mm	10 in.	1.78 mm (0.070 in.) total tolerance band or ±0.89 mm (±0.035 in.)	2.54 mm (0.100 in.) total tolerance band or ±1.27 mm (±0.050 in.)
508 mm	20 in.	3.05 mm (0.120 in.) total tolerance band or ±1.52 mm (±0.060 in.)	5.08 mm (0.200 in.) total tolerance band or ±2.54 mm (±0.100 in.)

(a) Improved tolerances may be possible depending on the specific foundry capabilities and overall part-specific requirements.

Table 10.   Surface finish of titanium castings

| Process | NAS 823 surface comparator(a) | rms equivalent (b) | |
		μm	μin.
Investment			
As-cast	C-12	3.2	125
Occasional areas of	C-25	6.3	250
Rammed graphite			
As-cast	C-30 to 40	7.5 - 10	300 - 400
Occasional areas of	C-50	12.5	500
Hand finished	C-12 to 25	3.2 - 6.3	125 - 250

(a) NAS, National Aerospace Standards. (b) rms, root mean square.

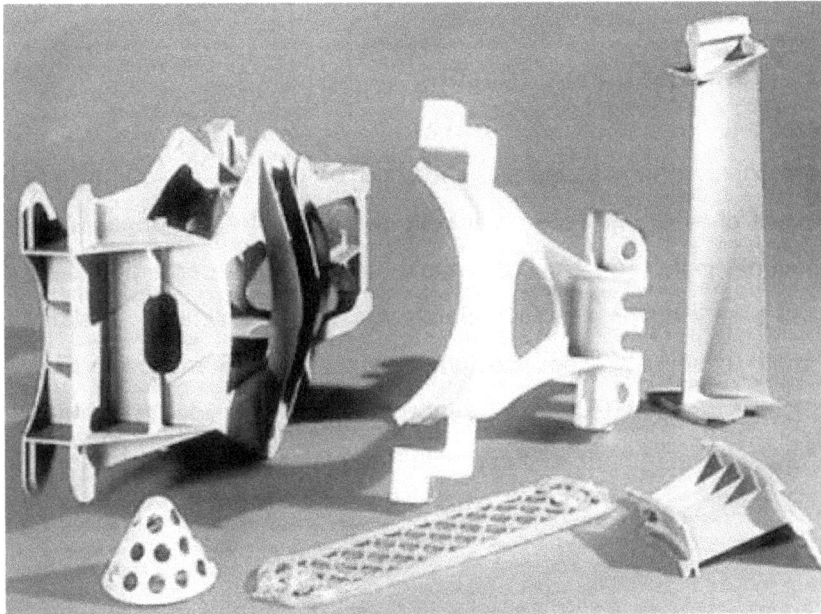

Figure 12. Investment cast titanium alloy airframe parts.

Figure 13. Typical investment cast titanium alloy components used for gas turbine applications.

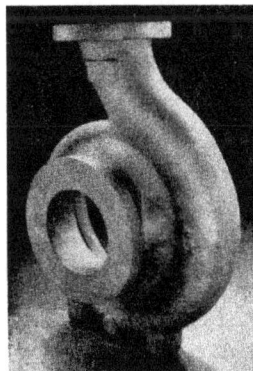

Figure 14. Titanium hydraulic housing produced by the investment casting process.

2.2. Permanent Mold Casting

Permanent mold casting is a method of casting in which a permanent mold (usually made of steel) is used repeatedly to cast metal components. Steps in the PMC process include mold construction, mold coating (in certain cases), casting, mold opening and part ejection, cut off, finishing, and inspection. Related issues include melting, mold design, process design, and mold maintenance. Figure 15[10] is a schematic of the steps in PMC. Reference 11 gives an in-depth description of the PMC process, which is summarized here with special attention given to titanium-specific issues.

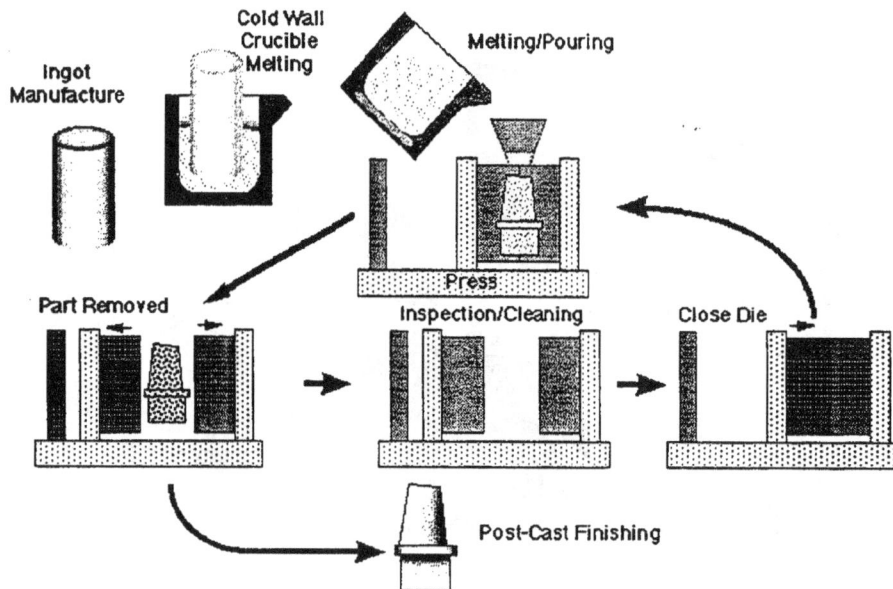

Figure 15. Schematic diagram of the permanent mold casting process.

2.2.1. Mold Design and Manufacture

The first step in PMC is mold design and manufacture. There are two basic designs for permanent molds, a book type, which opens and closes on a single hinge like a book, and an in-line type, which opens and closes along a single axis perpendicular to its parting line. The book type is the simplest, and is often used for manual PMC of simple shapes. The in-line type is used for deep castings and in automated casting machines.

The complexity of the part dictates the complexity of the mold. Simple parts can be produced with simple two-part mold sets. More complex parts can require multi-part molds, cores, and sliders. In many cases, mold inserts made of special materials or with special coatings are used in areas of possibly high wear. The use of chills and antichills (which promote directional solidification and allow for the feeding of solidification shrinkage) requires special passageways and further complicates the molds.

Because the molds are permanent (i.e., used repeatedly), mold design is very important and at times extremely difficult. The number of cavities in each mold set, the configuration and size of the gates and runners, and the number, placement, and size of vents and overflows must all be designed to minimize cost and maximize production. The placement of ejector pins or other ejection mechanisms and the use of replaceable cores and inserts must also be designed into the molds. Permanent mold castings must also be carefully designed to facilitate part removal. For example, a certain amount of draft must be left and drastic changes in thickness should be avoided.

Before the mold can be constructed, a mold material must be chosen. Material choice depends on the alloy being cast (melting temperature and reactivity), the size of the parts being cast, the number of cavities per mold, the cost of the mold materials themselves, and the cost of mold fabrication (machining, heat treating, etc.) as it relates to specific mold materials. The choice of mold materials will be addressed in more detail in the mold wear section of this report.

Once the mold material is chosen, a manufacturing method must be chosen. Molds can be cast to shape or can be machined with conventional CNC methods or with EDM. Surface finish of the mold cavity is directly related to surface finish of the casting, and should be specified accordingly. Molds for castings which require the best surface finishes are ground and polished before they are put into service.

Once a mold is fabricated, special coatings are sometimes applied to protect it from the molten metal. Coatings can be used to modify the heat transfer properties of the mold or to lubricate it to prevent soldering. Coatings can be in the form of slurries, suspensions, or liquid solutions and can be sprayed or brushed on. Insulating coatings must be applied before lubricating coatings to achieve proper results. The thickness and surface finish of the coating affect the properties of the casting. The thicker the coating,

31

the better it insulates, but the more likely it is to spall off. The smoother the surface of the coating, the smoother the casting, but the more likely oxide skin formation and the trapping of gases becomes. The coating must be thoroughly dry before casting begins to prevent the generation of gases, which can lead to increased porosity. Coating life varies with process parameters. Most insulating coatings do not need to be reapplied after each casting cycle, but many lubricating coatings do.

2.2.2. Casting and Finishing

The casting cycle in PMC can be manual or automated, but the basic steps are similar. First, the mold is closed and locked. Then the molten metal to be cast is poured into the mold cavity under the influence of gravity. The mold remains closed for a certain amount of time to allow the component to solidify, after which the mold is opened and the casting removed. The mold is then prepared for another cycle by cleaning and adding any necessary lubricants. New insulating coatings are applied when required. The mold is closed and locked, and a new cycle begins. This casting cycle is repeated again and again until the mold starts producing unacceptable castings. At this juncture, the mold may be repaired by welding and grinding or may be taken out of service permanently and replaced.

Once the casting is removed, the gating and flash are trimmed away manually or with a trim press. After inspection for gross defects, the trimmed parts are ready for HIP, weld repair, and heat treatment. Final machining, stress relieving, and strengthening operations are performed, and the castings are inspected. Inspection mechanisms are similar to those used for investment castings.

2.2.3. Mold Wear Mechanisms/Mold Wear Observations in PMC

Mold wear can be a major problem in permanent mold casting and die casting. Wear of the mold surfaces can cause castings to lose dimensional accuracy, to have poor surface finish, or to stick to the molds. As molds wear, they must be removed from service for repairs. Eventually, mold wear can cause catastrophic failure of the molds. Both repair and failure of molds are costly due to the high cost of mold fabrication and due to the loss of production experienced when molds are removed from service. Because mold wear can be so costly, producers attempt to minimize mold wear by proper casting design and process control as well as by using mold coatings or lubricants.

Mold wear and mold failure in casting processes occur via several mechanisms- erosion, abrasion, adhesion, corrosion, and thermal fatigue (heat checking)- which can operate separately or together to shorten mold life.[12] Upon mold filling, molten metal impinges on the mold surface causing erosion. If the molten metal contains solid particles, mold wear by abrasion occurs as the metal flows past the mold. As the molten metal remains in contact with the eroded and abraded mold, chemical reaction can take place, and the mold can be corroded by the molten metal. When solidification starts, the solidifying metal can solder to the mold, which can in turn cause damage to the mold surface and the casting upon ejection. Throughout the casting cycle, the molds are subjected to large thermal gradients. These gradients cause the buildup of internal stresses due to thermal expansion. The cycling of these internal stresses can cause fatigue cracks to initiate and grow on the mold surface. This thermal stress cycling is the mechanism of heat check formation, and can lead to poor surface finish and ejection difficulty.

By carefully designing the gating system and fluid flow patterns in mold filling, mold wear by erosion can be minimized. Molten metal impingement on movable cores, protrusions, or corners should be avoided. Additionally, mold inserts should be used to enhance wear resistance in critical areas. Mold inserts are usually made of a more expensive, wear-resistant material; they extend mold life because they wear more slowly or can be replaced if they become overly worn.

Corrosion of the mold by molten metal can be lessened by carefully controlling the temperature of the melt and by coating the mold with corrosion-resistant material. As corrosion accelerates with increasing temperature, pouring at the minimum possible melt temperature is better, but considerations of fluidity must be taken into account. Mold coatings are useful for adding to the life of a mold, but the additional cost and effect on casting properties such as surface finish must be considered.

Lubrication can aid in the ease of ejection of castings from molds by preventing soldering. The choice and use of lubricant is critical. Lubricants that react with the molten alloy can adversely affect casting quality by forming brittle surface layers. Lubricants that corrode the mold must be avoided. Liquid lubricants that evolve gas upon heating can cause porosity near the casting surface. Too much lubricant or lack of binding to the surface of the mold can cause areas of poor filling. The use of too little lubricant can result in soldering.

Mold design and process design are important for controlling heat checking. Choice of a mold material that resists thermal fatigue is important. Materials with low coefficients of thermal expansion and high thermal conductivity are desired for resistance to heat checking. Minimization of thermal gradients throughout the process also helps to minimize heat checking. Designing the mold cavity to minimize large changes in cross-section can help to minimize thermal gradients within the mold. Eliminating sharp corners or small radii helps minimize stress concentrations. Preheating the mold also decreases thermal gradients, and die casting molds are typically run as hot as possible.

2.2.4. PMC Defects

Common PMC defects are porosity, inclusions, dross, cold shuts and misruns, poor surface finish, and distortion and cracking. Most of these defects can be controlled by proper mold design and process control. Porosity can be caused by trapped gas, insufficient feeding of molten metal to thick sections (macroporosity), or by insufficient cooling rates for wide-freezing-range alloys (microporosity). Inclusions and dross result from "dirty" molten charges and molds. Cold shuts and misruns occur when the molten alloy is not sufficiently fluid to completely fill the mold, and can be avoided by adding more superheat to the melt, by using centrifugal casting, or by preheating the mold. Poor surface finish results when soldering occurs, when there is surface connected porosity, when two streams of alloy freeze as they impinge, thus leaving a trace of their boundary on the surface of the casting (seams), or when the surface finishes of the molds are poor. Distortion and cracking result when the casting is overly constrained by the mold as it cools (and contracts) or when uneven cooling causes large thermal gradients in the casting.

Besides these defects, permanent mold castings may have variable dimensional accuracy. As with any production process, there are cyclic variations and trend variations that affect the process. From casting to casting, dimensional variations can be caused by improper mating of the molds or variations in cycle time. As casting continues, mold wear and mold distortion begin, and casting dimensions lose tolerance. Excessive flash can form and heat checks on the mold surface can harm surface finish of the castings.

2.2.5. Permanent Mold Casting of Titanium Alloys

Permanent mold casting of titanium alloys is currently in the development stages. Several aerospace and automotive components have been permanent mold cast on a laboratory scale, and evaluation of such components is currently underway. Aerospace components such as Ti-6Al-2Sn-4Zr-2Mo bell crank supports, Ti-6Al-4V linkage brackets, Alloy C+ variable vanes, Ti-6Al-4V hollow turbine blades for military aircraft, and titanium aluminide automotive valves have been produced via PMC.[13],[14] Figure 16[10] shows some representative PMC titanium parts. Preliminary mechanical testing of the aerospace components shows that the PMC components meet Pratt & Whitney Aircraft (PWA) minimum specifications for UTS, yield strength, elongation, reduction in area, and fatigue strength. These favorable test results have induced Pratt & Whitney to add PMC as an alternative to investment casting in several of their materials specifications.[14]

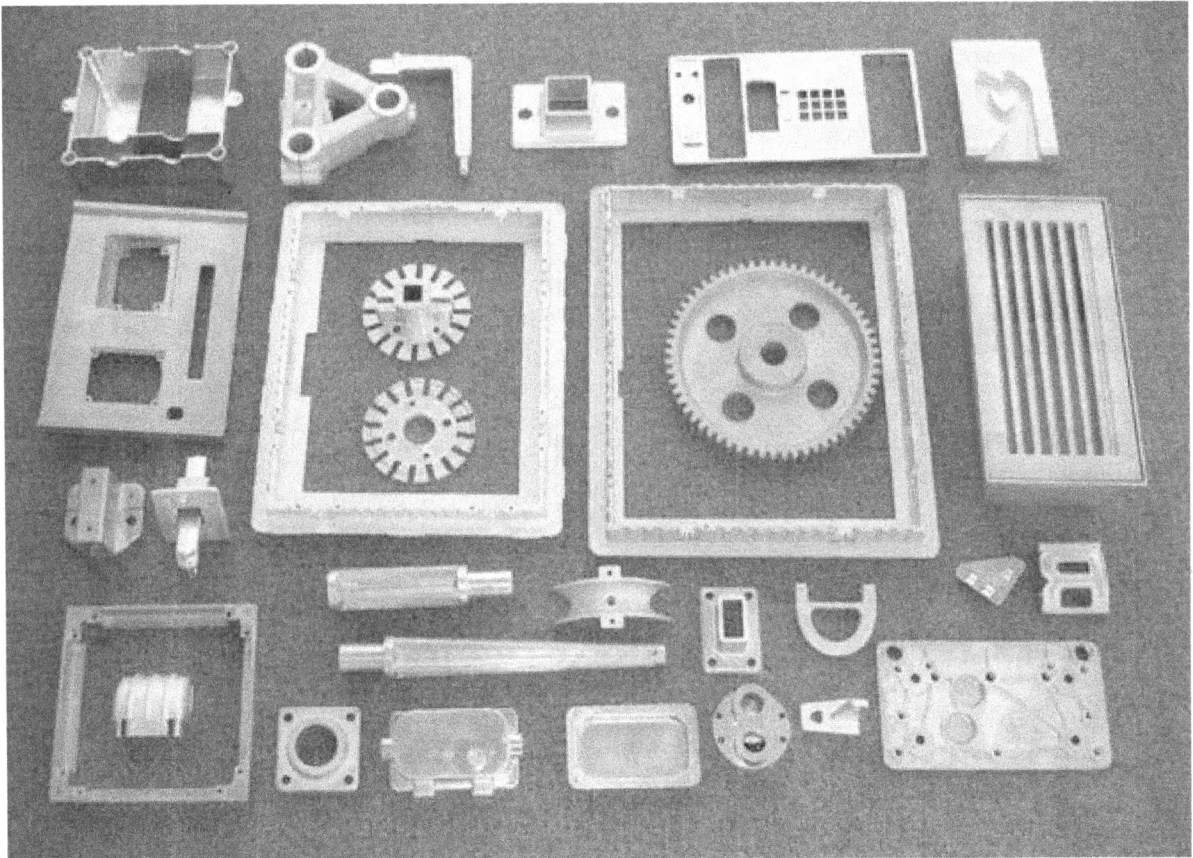

Figure 16. Titanium components produced using the PMC process.

Although PMC has been shown to be capable of producing acceptable Ti alloy components, the process is by no means perfected. For example, during the preliminary casting trials, difficulty was encountered with mold wear in sharp radii and with component distortion during the HIP cycle.[13],[14] The mold wear in sharp radii appeared to be caused by mold-metal reaction, as many of the permanent mold cast components exhibited areas of iron enrichment near sharp radii (Figure 17[14]). Investigators believe a local increase in temperature causes mold overheating in sharp corners, which, in turn, increases the amount of chemical interaction between the molten alloy and the mold. Distortion during the HIP cycle is believed to occur when great amounts of off-center porosity exist. Future work in the area of mold wear in Ti PMC is desired before full-scale production begins. Determination of the limits that mold wear puts on cavity shapes is essential if PMC is to be seriously considered as a manufacturing method. The ability to predict mold life is required to make accurate cost projections. Studies on the effect of coatings and lubricants on wear of Ti PMC molds as well as studies on metal-mold reaction for different mold materials could lead to improvements in mold life or to the ability to PMC titanium components in finer detail.

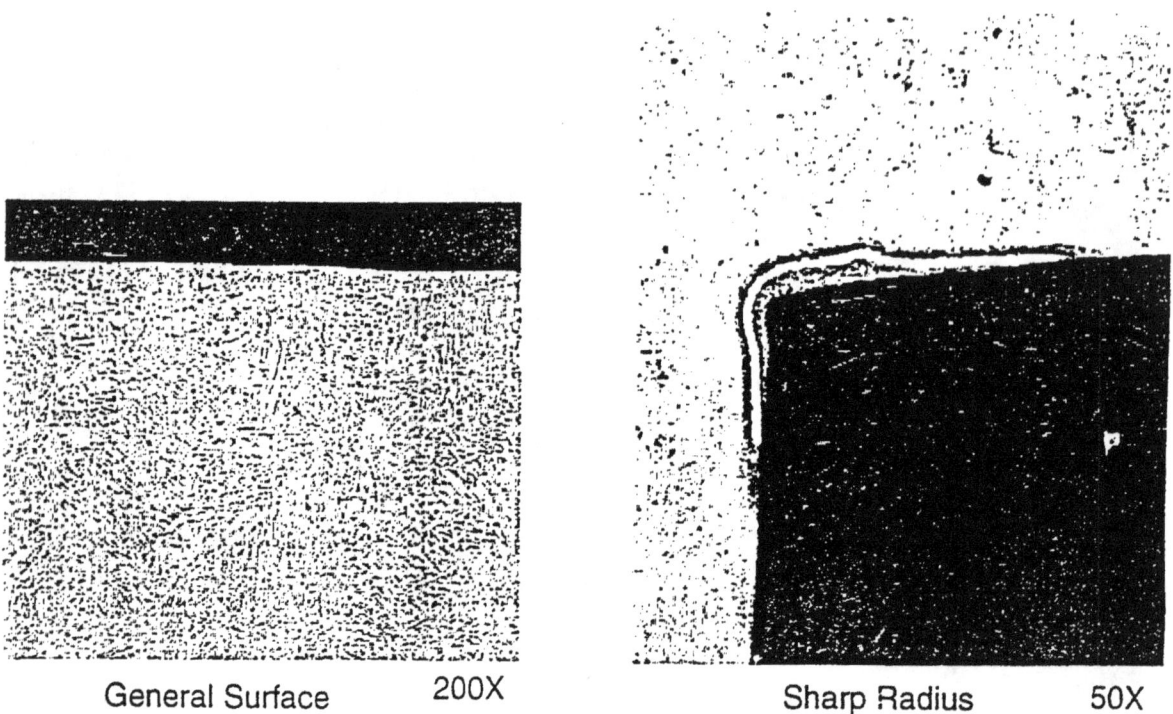

General Surface   200X   Sharp Radius   50X

Figure 17. Representative microstructure for a certain permanent mold cast titanium component showing surface reaction in the sharp fillet radius.

Permanent mold casting could prove to be a cost-effective manufacturing method for titanium components. The main advantage of PMC over investment casting is the elimination of processing steps (Figure 18[15]), which reduces cost and allows for tighter process control. Another advantage of PMC of titanium alloys is the lack of alpha case formation. Because no alpha case forms, parts can be cast to closer tolerance and finishing costs can be significantly reduced. The greater cooling rates seen in PMC relative to investment casting also result in castings with a finer microstructure, which can also be advantageous in terms of mechanical properties. Figure 19[15] shows both PMC and investment cast microstructures. Provided that mold wear does not become cost-inhibiting, PMC could become a viable alternative to investment casting for many titanium alloy components.

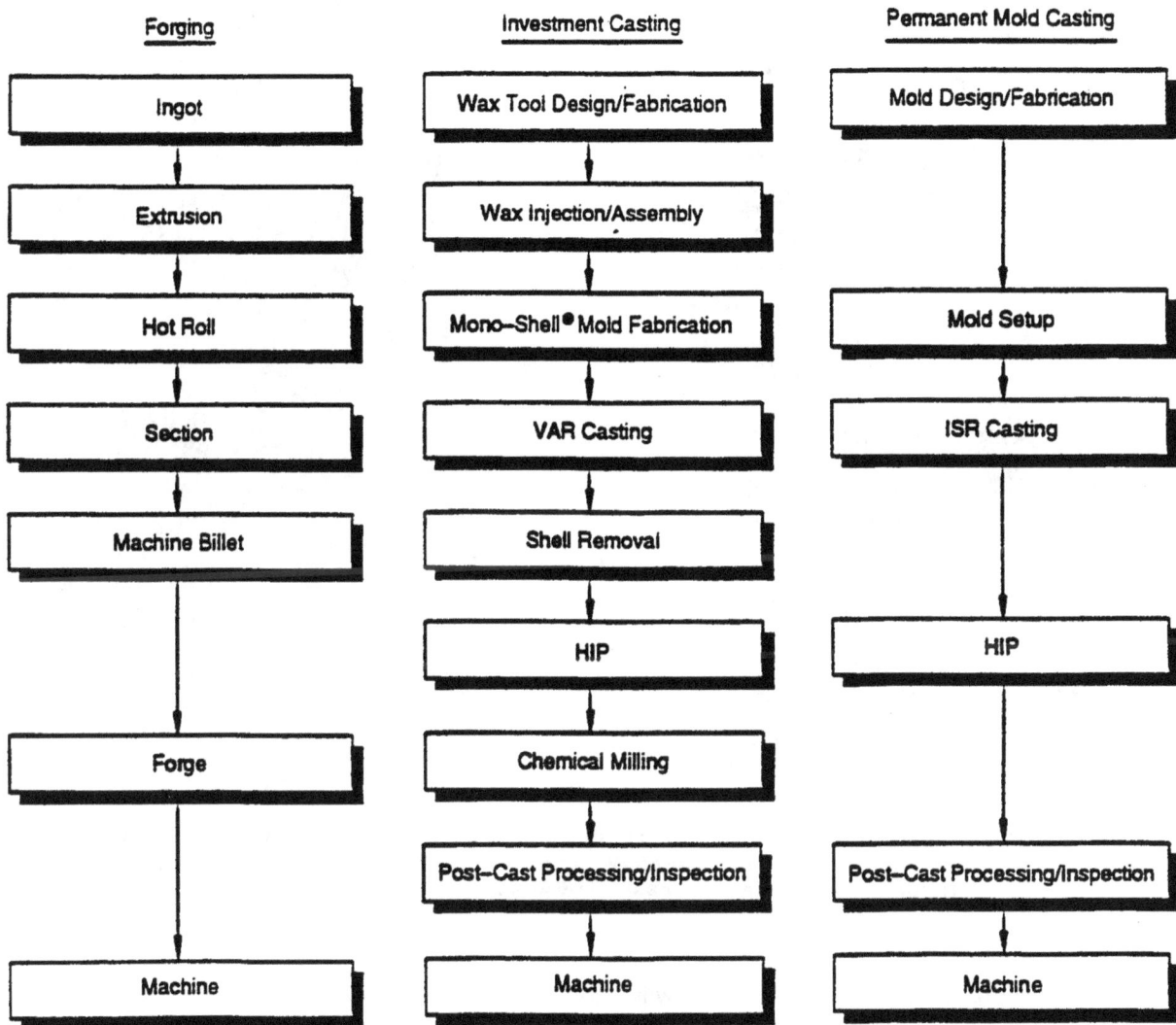

Forging	Investment Casting	Permanent Mold Casting
Ingot	Wax Tool Design/Fabrication	Mold Design/Fabrication
Extrusion	Wax Injection/Assembly	
Hot Roll	Mono-Shell® Mold Fabrication	Mold Setup
Section	VAR Casting	ISR Casting
Machine Billet	Shell Removal	
	HIP	HIP
Forge	Chemical Milling	
	Post-Cast Processing/Inspection	Post-Cast Processing/Inspection
Machine	Machine	Machine

Figure 18. Comparison of the number of steps in forging, investment casting, and permanent mold casting.

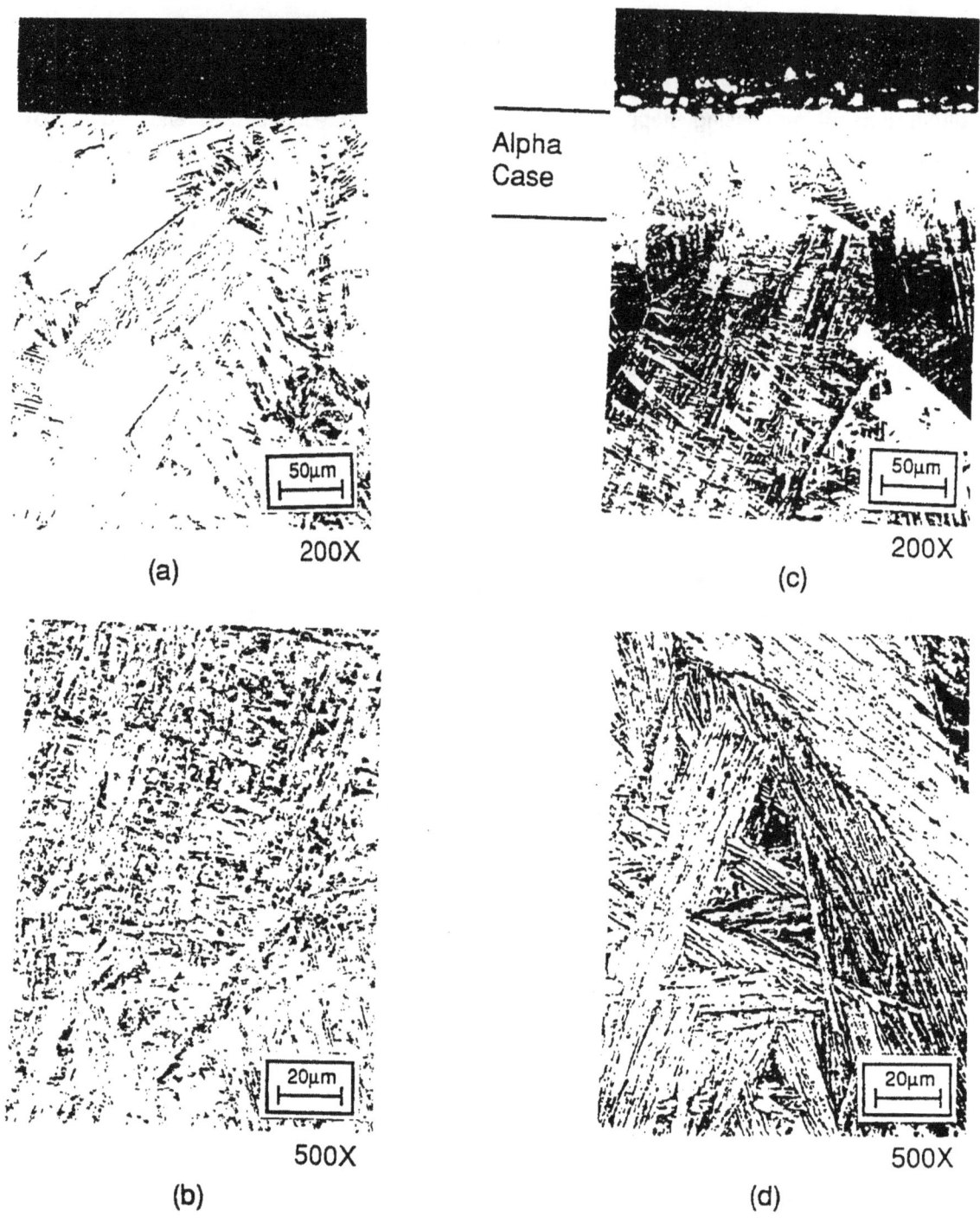

Figure 19. Typical microstructure (a) at the surface, and (b) in the interior of permanent mold cast Ti-6Al-4V, and (c) at the surface, and (d) in the interior of investment cast Ti-6Al-4V.

Work in PMC of Ti alloys will continue, with future goals being Ti alloy PMC/solidification modeling for mold design and process optimization, greater understanding of mold wear issues, alloy development for PMC, determination of manufacturing efficiency, and eventually scale-up to production level. Little work has been done to model PMC of titanium alloys, and heat transfer coefficients have not been adequately determined. All analysis of mold wear has been qualitative, and the molds used in casting trials have not been used for a large number of pours, thus little is known about mold life.

2.3. Discussion of Research Issues/Approaches

Research issues in casting of titanium alloys include:

- Determining the castability of specific titanium alloys
- Formulating special casting alloys
- Predicting mold life for specific casting systems
- Formulating special mold materials, inserts, coatings and lubricants
- Determining the feasibility of die casting (pressure injection)

Castability studies would reveal the geometric capabilities of the PMC process, such as minimum achievable cross-section size and corner radius, thus providing guidelines for designers. Some simple experiments using traditional castability molds and fluidity spirals could be conducted to obtain preliminary data. Follow-up experiments using specially designed molds could be conducted to obtain further data.

Development of special casting alloys that improve mold fill or surface finish or decrease porosity would enhance the capabilities of PMC, thus increasing the number of components that can be produced via PMC. Mold fill comparisons between alloys could be facilitated by fluidity spiral tests. Surface finish and porosity tests would simply involve the inspection of various castings.

Experiments that compare the performance of different mold materials on the basis of mold wear could help mold designers choose the optimum mold material based on both performance and cost. The development and evaluation of special mold inserts, coatings, and lubricants would further aid the progress of the titanium PMC industry.

3. Modeling of Microstructure Evolution During Casting

The mechanical properties of castings strongly depend on microstructure, which in turn strongly depends on the thermal conditions during casting. For stressed components, casting conditions must be controlled to ensure adequate mechanical properties. Without modeling, casters have to determine the proper heat removal characteristics by trial and error or experience-based guesses. Trial and error can prove to be very costly, thus techniques to predict microstructure based on measurable thermal parameters are attractive. The basis for such methods is described in detail in the following sections. Section A describes experimental methods used to determine relationships between microstructural features and thermal parameters; section B gives examples of the use of these methods to predict structure and defects in non-titanium alloys. Section C describes analytical modeling of microstructure evolution during solidification. Sections D and E contain summaries of microstructure observations for titanium castings and titanium powders, respectively. Finally, section F describes research issues and approaches for modeling microstructure evolution during solidification.

3.1. Construction of Solidification (G vs. R) Maps

A large amount of research has established that microstructural features and casting defects can often be correlated to the solidification rate and the thermal gradient at the solid-liquid interface. The liquid alloy composition, nuclei density in the liquid, fluid flow conditions near the liquid-solid interface, and local solidification time also affect microstructural features of castings.

Solidification maps (also referred to as microstructure maps) are used to graphically depict the relationship between casting conditions and structural features. These maps have solidification rate, R, on the x axis and thermal gradient, G, on the y axis. Cooling rate, dT/dt, is related to R and G by the relationship $dT/dt = -G \cdot R$. Experimentally determined relationships called "criterion functions" relating G and R to microstructural features are plotted on the solidification maps for specific values of all other variables. Relationships for the columnar to equiaxed transition, secondary dendrite arm spacing, eutectic structure, and interdendritic porosity and other defects exist for certain alloys.

Once all the pertinent criterion functions are plotted, an "operating window" can be determined for the casting process. The operating window consists of a range of G and R values for which the casting most likely will have the desired microstructure and properties (Figure 20[16]). Because the plotted functions are empirical and tend to be "order-of-magnitude" type relationships, the operating window usually only serves as a guide, and cannot be viewed as an absolute calculation. Minor adjustments of an initial guess are often required to obtain optimal castings.

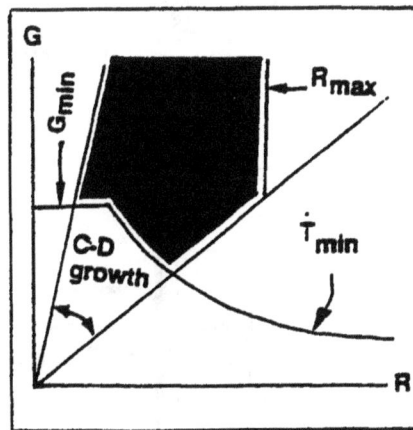

Figure 20. An example solidification map showing the process target zone.

Empirical determination of various criterion functions to construct the solidification map and evaluation of the validity of the solidification map are done experimentally. Special pieces of apparatus in which G and R can be controlled and easily determined are used to make simple castings. These castings are sectioned and metallographically prepared to determine macrostructure, secondary dendrite arm spacing, the number of effective nucleation sites, and other microstructural features. For complex castings, values of G and R are computed from solidification modeling, and the metallurgical data are plotted on the solidification map. These results are compared to the predictions of the criterion functions to determine the validity of the map.

For low values of G and R, a Bridgman-type unidirectional solidification apparatus is used (Figure 21[17]). In this system, R and G are set by fixing the mold removal rate and the furnace set temperature. A thermocoupled mold is used with a data acquisition system to record the temperature history of the alloy at certain positions. The thermal gradient is calculated from the thermocouple data and the solidification rate.

41

Unthermocoupled molds are then used under the same conditions to produce a specimen for metallographical inspection.

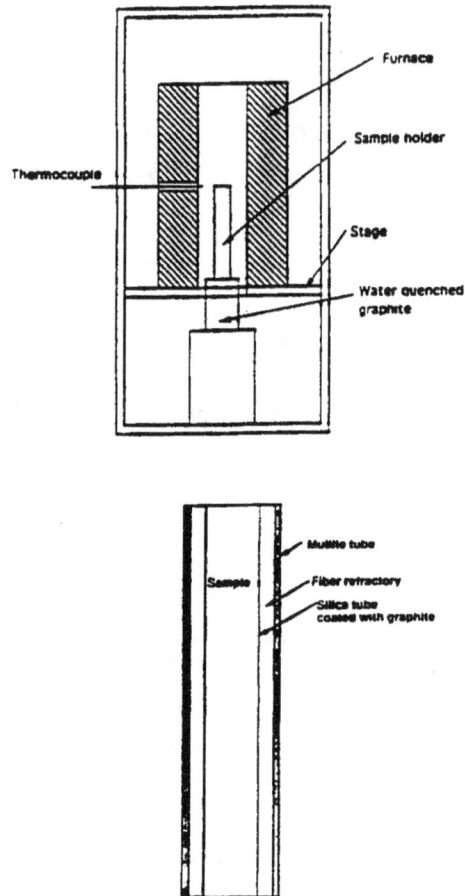

Figure 21. Schematic diagrams of a) the Bridgman-type unidirectional solidification apparatus and b) the sample holder used in the apparatus.

Larger cooling rates can be achieved using a unidirectional heat removal mold or a three dimensional heat removal mold. An example of a unidirectional heat removal mold is one constructed of graphite and insulated with a castable refractory such as that developed by Tsumagari and Mobley (Figure 22).[17] The bottom of the mold is cooled with a water jet and the top of the mold is heated with a torch. Thermocouples are used to determine the temperature history of the casting, from which R and G values can be easily calculated. Again, the castings are sectioned and metallographically inspected, and the results are used to develop the solidification maps.

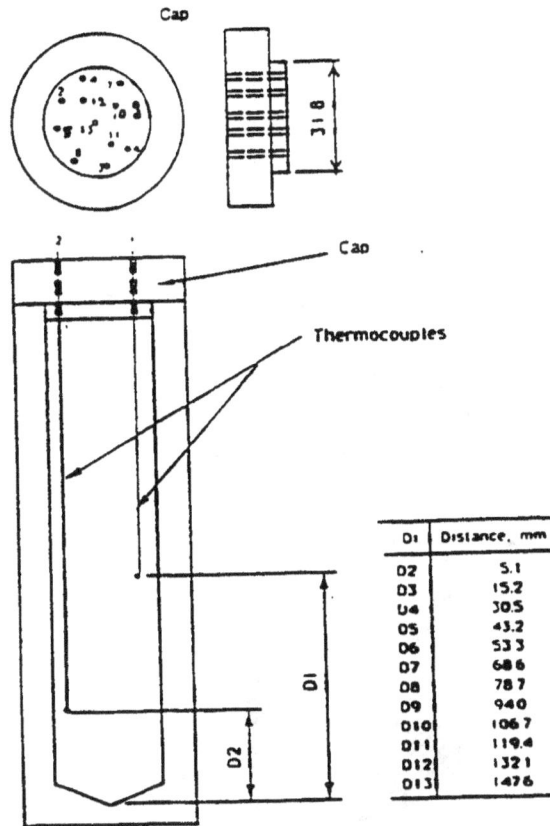

Figure 22. Schematic diagram of the unidirectional heat removal mold.

An example of a three dimensional heat removal mold is a simple cooling cup, which does not include any extraneous heating or cooling (Figure 23).[17] Thermocouples are used to measure the temperature history of the casting, but solidification modeling must be used to determine G and R. This complicates the experiment because modeling requires input such as thermophysical properties and interface heat transfer coefficients. Provided that a good fit between the experimental and modeled temperature profiles exists, the resultant G and R can be correlated with measured microstructural features to determine a solidification map.

Once the solidification map is confirmed, it can be used in practice to predict microstructural features of actual castings. Casting conditions such as superheat, initial mold temperature, environment temperature, and mold and casting geometry in conjunction with casting and mold thermophysical properties and interface heat transfer

43

properties can be input into a solidification modeling package to determine time-averaged values of G and R for specific locations within the casting. Casting thermal parameters can then be modified to achieve the desired microstructure via control of G and R. Solidification maps therefore lead directly to the increased usefulness of solidification modeling.

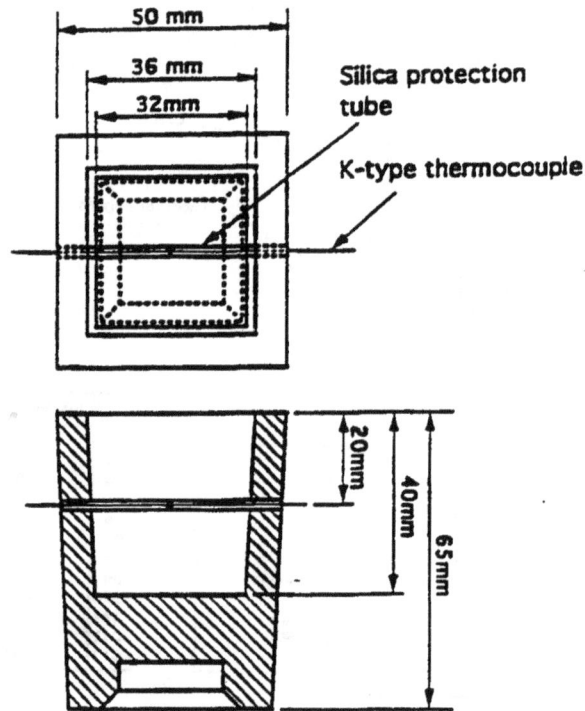

Figure 23. Drawing of a cooling analysis cup made of resin-bonded sand.

Despite the appeal of solidification maps, caution must be exercised in their utilization. The empirical nature of the critical parameters plotted on solidification maps, and the disregard of nucleation and growth kinetics, geometric effects, and time-varying G and R emphasize the fact that these maps are guidelines depicting trends and are therefore not infallible. As solidification modeling advances to include more complete descriptions of the physics of solidification, more accurate values of G and R and better-fitting criterion functions will result. The accuracy of solidification maps for predicting the microstructures and properties of actual castings will therefore improve.

3.2. Solidification (G vs. R) Map Examples

Many examples of the construction and application of solidification maps can be found in the literature. Tsumagari and Mobley[17] constructed solidification maps for A356 and D357 aluminum alloys, Overfelt[18] for superalloys, and Yu et al.[19], Tu and Foran[20], and Purvis, Hanslits, and Diehm[21] for single crystal investment castings. Examples from their work are presented below.

3.2.1. Aluminum Alloys A356 and D357

Tsumagari and Mobley related features such as grain size, secondary dendrite spacing, dendrite morphology, interdendritic porosity, and eutectic morphology and spacing to G and R through the use of solidification maps for two aluminum alloys- A356 and D357. A description of their methods follows.

The selected aluminum alloys (A356 and D357) have two major solidification products, primary alpha Al and an Al-Si eutectic; therefore two different solidification maps were needed. The first map was for the primary alpha Al phase; it related the grain size, secondary dendrite arm spacing, alpha dendrite morphology, and percent interdendritic porosity to G and R. The second map was for the eutectic phase; it related the morphology and spacing of the eutectic aluminum-silicon structure to G and R.

For the primary alpha map, Hunt's criterion for the columnar-to-equiaxed transition[22] was applied. The map was thereby divided into three distinct regions - a fully columnar region (in which $G \geq 2.86(N_0)^{1/3}[1-(\Delta T_n/\Delta T_c)^3]\Delta T_c$, N_0 = the number of effective nuclei, $\Delta T_c = \{-8\Gamma m_l(1-k)C_0R/D\}^{1/2}$, ΔT_n = undercooling required for interface growth, Γ = Gibbs-Thompson coefficient, $\sigma/\Delta S_f$, σ = liquid-solid interfacial energy, ΔS_f, = entropy of fusion, m_l = liquidus slope, k = equilibrium partition ratio, and D = diffusivity in the liquid), a fully equiaxed region (in which $G \leq 0.617(N_0)^{1/3}[1-(\Delta T_n/\Delta T_c)^3]\Delta T_c$), and a mixed region (in which $2.86(N_0)^{1/3}[1-(\Delta T_n/\Delta T_c)^3]\Delta T_c \leq G \leq 0.617(N_0)^{1/3}[1-(\Delta T_n/\Delta T_c)^3]\Delta T_c$). Most of the parameters in Hunt's equations were obtained from the literature[23], but proper values of N_0 and C_0 had to be chosen for each alloy and each casting configuration. In this case, N_0 was varied from 1 to 10,000 /cm^3 and C_0 was varied from 6.5 to 7.5 %Si to show the range of possibilities.

Next, lines of constant dT/dt were plotted based on the relationship dT/dt = -G•R. These lines of constant dT/dt were correlated to lines of constant secondary dendrite arm spacing (SDAS = 35 (dT/dt)$^{-1/3}$, as reported in Spear and Gardner[24], with dT/dt in units of K/s) and percent interdendritic porosity (%P = 0.26 (dT/dt)$^{-0.91}$, as reported in Fang and Granger[25]). The resulting map for a single value of N_0 is shown in Figure 24.[17]

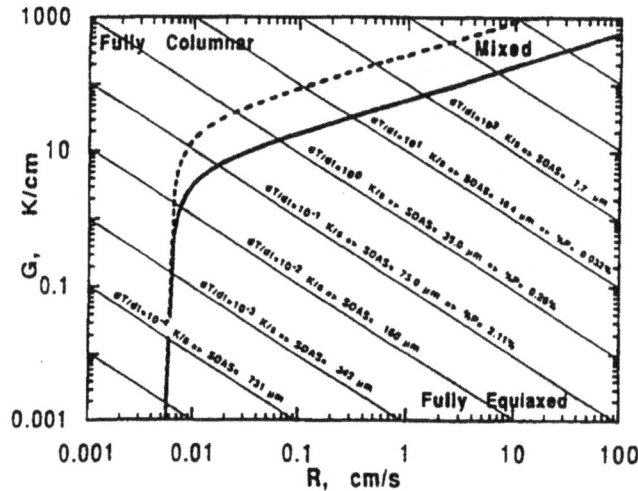

Figure 24. Solidification map for α-Al primary phase in A356 and D357 alloys. The morphology of dendrite, grain size, secondary arm spacing, and percent of interdendritic porosity are predicted from this map.

For the eutectic solidification map, the important transition is from a flake-like to rod-like eutectic silicon morphology. The rod-like morphology is usually preferred because it increases ductility. The critical cooling rate for the transition from a flake-like to a rod-like structure is reported to be ~10 K/s.[26] This line of constant dT/dt was plotted on the solidification map. Secondly, the eutectic spacing is given by $\lambda^2 R$ = 3.87 x 10^{-7} mm^2/s, in which λ is the average silicon eutectic spacing.[27],[28] This results in vertical lines of constant spacing corresponding to the given value of R. The resulting map for the Al-Si eutectic is shown in Figure 25.[17] Because the scales on both the primary and eutectic maps are identical, the two maps may be superimposed.

Once the maps were constructed, validation experiments were conducted. These comprised the Bridgman-, the unidirectional heat removal-, and the three-dimensional heat removal types. R and G were determined from thermal histories and numerical modeling techniques. The castings were sectioned and metallographically examined. The grain density (i.e., number of nucleation sites), dendrite morphology, secondary dendrite arm spacing, and cooling rate were determined. The results from the Bridgman and unidirectional experiments were plotted on the alpha - Al solidification map, and were found to correlate well with the predicted trends (Figure 26[17]).

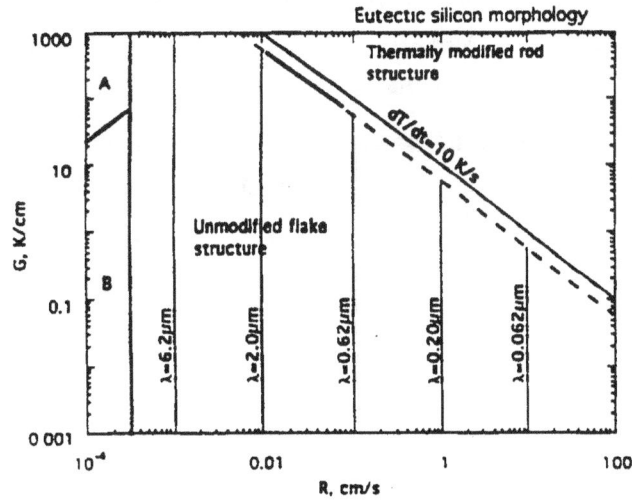

Figure 25. Solidification map for the Al-Si eutectic structure in A356 and D357 alloys.   In the region A, the eutectic silicon has a (100) texture.   In the region B, the structure is mixed between the Si with (100) texture and the flake-like Si, which contains multiple {111} twins. The symbol λ represents the average spacing of the Al-Si eutectic.

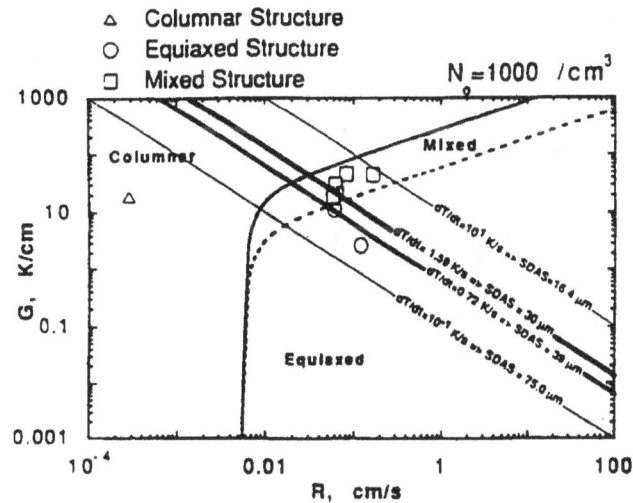

Figure 26. Solidification map for the primary α-Al phase showing experimentally obtained R-G plots.   A triangle point represents the columnar structure obtained from the Bridgman-type furnace, and square and circular points represent equiaxed and mixed structures, respectively, obtained from the unidirectional heat removal mold experiment.   The R-G values in the cooling analysis cup samples were predicted to be located in the region divided by two thick lines.

47

3.2.2. Ni-Base Superalloys:

Overfelt constructed similar maps for nickel-based superalloys based on the data of McLean[29] and Bouse and Mihalisin.[30] In Figure 27a[18], lines of constant $R^{-0.25}G^{-0.5}$ corresponding to constant λ_1 (primary dendrite arm spacing) were plotted, while in Figure 27b[18], lines of constant $G \cdot R$ corresponding to constant λ_2 (secondary dendrite arm spacing) were plotted. The first map was constructed by fitting experimentally determined data to the relationship $\lambda_1 = KR^{-0.25}G^{-0.5}$ to determine the value of the alloy dependent parameter K. The second was constructed by fitting experimentally determined data to the relationship $\lambda_2 = C(G \cdot R)^n$, in which C is an alloy dependent constant and $1/3 \leq n \leq 1/2$. The data were obtained from precision directional solidification experiments.

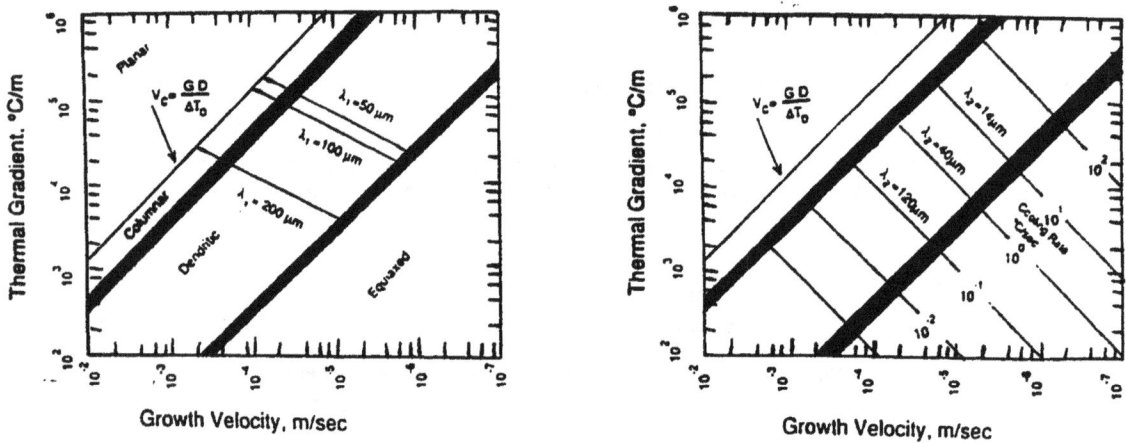

Figure 27. **Microstructure maps showing the morphology and dendrite-arm spacings for nickel-based superalloys. Shown is the critical velocity for plane-front growth. (a) λ_1 data from McLean. (b) λ_2 data from Bouse and Mihalisin.**

Yu et al. constructed a solidification map for single crystal investment cast superalloys. Investment casting trials were run on thermocoupled clusters of cored cylinders of various dimensions at different furnace temperatures and withdrawal rates. The castings were metallographically prepared and inspected to determine microstructural features and size and distribution of defects. The casting process was modeled using FEM, and the model was verified with thermocouple measurements. Solidification conditions (G,R) were calculated from the modeling results for each casting configuration. After the casting and modeling were completed, attempts were made to correlate the measured microstructural features and defects to the casting conditions. First, it was assumed that a critical local solidification time, t_f^*, exists below which no detectable freckle defects will form and that this critical time corresponds to a critical cooling rate (GxR) above which freckle defects will not form. The freckle criterion function is then GxR = ΔT/t_f^*. The critical GxR value for freckle formation was determined from the casting and modeling results and was plotted on the G versus R map. Secondly, the critical condition for equiaxed grain formation was postulated to be that G/R must be lower than a certain critical value. Again, this critical value was determined from the casting and modeling results and plotted on the G versus R map.

Finally, a parameter relating microporosity to casting conditions was plotted. After considering Niyama's shrink criterion function, $\sqrt{(G/R)}$,[31] and Lecompte-Beckers' microporosity index[32] for directionally solidified castings, $\Delta p = \{[24\mu\beta'n\tau^3(T_l-T_s)]/[\rho_l g]\}\bullet(R/G)$, in which m is the liquid viscosity, $\beta' = (\rho_s-\rho_l)/\rho_l'$, ρ_l is the liquid density, ρ_s is the solid density, n is the number of interdendritic channels per unit area (related to primary dendrite arm spacing), τ is the tortuosity (related to secondary dendrite arm spacing), T_l is the liquidus temperature, T_s is the solidus temperature, and g is the gravitational constant, the researchers chose G/R close to the solidus temperature as the critical function relating to shrinkage. In this case, higher values of G/R close to the solidus temperature (i.e., near the end of solidification, which is when porosity forms) were postulated to be more likely to produce sound castings. This fact was verified by the casting and modeling results, and the critical value of G/R for microporosity formation was plotted on the G versus R map. The resulting solidification map is shown in Figure 28[19] with arrows indicating the direction of increased defect formation tendency.

Tu and Foran also constructed maps to determine operating windows for single crystal investment castings. Solidification parameters were postulated for the equiaxed-to-columnar transition (G/R, as suggested by McLean[29]), freckle formation (G$\bullet$R), and microshrinkage (G/R, as suggested by Niyama[31]). Then directional solidification experiments were performed to determine the critical values of these parameters. The experimental procedure used to determine the critical parameters was straightforward. Many identical castings (single thermocoupled bars) were made in a small directional solidification furnace varying only the withdrawal rate. First a very slow withdrawal rate casting and a very fast withdrawal rate casting were made with the intent of producing an equiaxed casting and a columnar casting. The withdrawal rates between two castings of different type (i.e., equiaxed and columnar) were averaged to determine the withdrawal rate for the next casting. This process of averaging was repeated until the critical withdrawal rate for the transition was narrowed down to a small range of withdrawal rates.

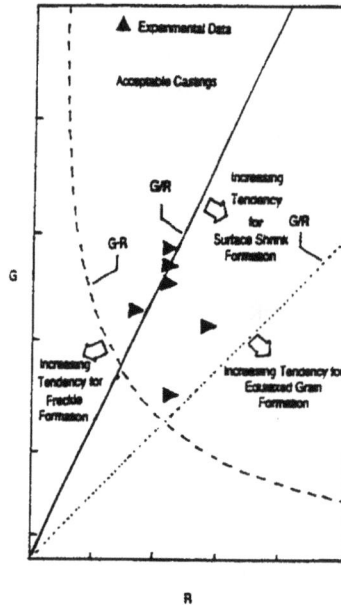

Figure 28. An example solidification map for single-crystal investment cast superalloys.

Once the critical withdrawal rate was determined, the process was modeled to calculate the values of G and R and the model was validated with the measured data. To this end, the critical value of the appropriate criterion function (i.e., G/R) was determined by computing the average value of G and the average value of R for the entire casting and inserting these average values into the proper function. Finally the critical value line was plotted on the defect map and the corresponding areas were labeled.

To validate the defect map, experiments on an actual cluster of single crystal turbine blades were performed. The entire cluster was modeled using ProCAST and validated with thermocouple measurements. G and R values were computed for each FEM node of the casting, and each set of computed G and R values were plotted on the defect map. The results can be seen in Figure 29.[20] This particular casting showed a slight tendency for equiaxed grain formation and no tendency for freckle formation based on the defect map, but no discussion of the actual structure of the test casting was presented.

50

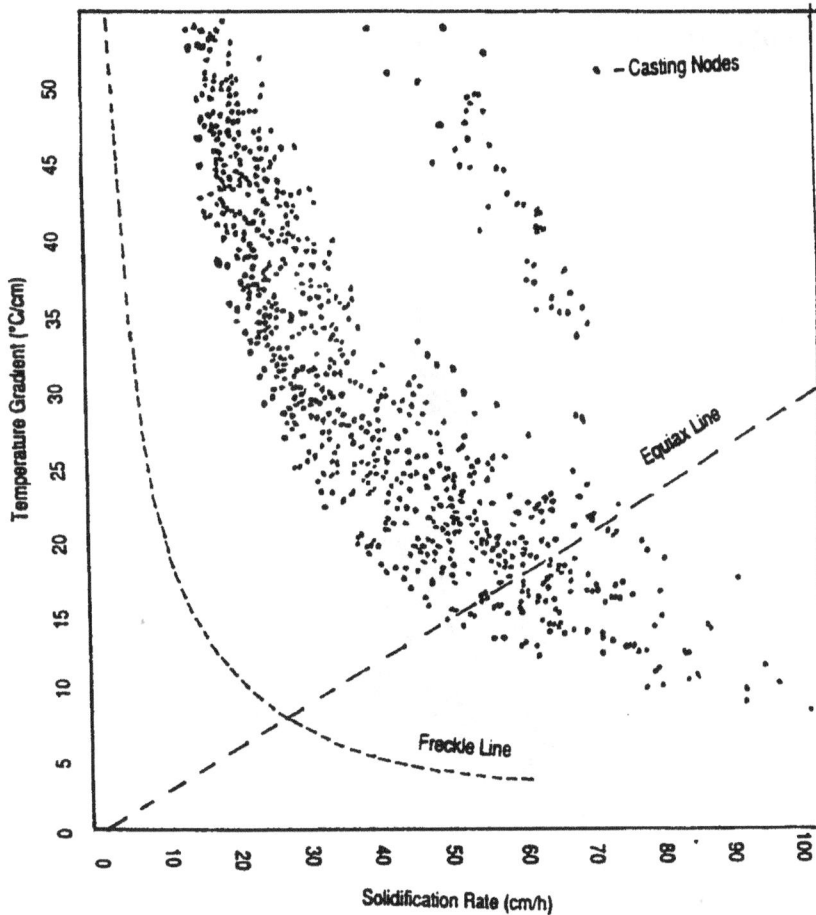

Figure 29. The solidification map for a single-crystal investment cast superalloy turbine blade.

Purvis, Hanslits, and Diehm conducted microstructure modeling experiments on a special single-crystal superalloy investment casting. Figure 30[21] shows the geometry of the test casting, which was designed to include several changes in cross section to show how geometry affects microstructure. The casting was modeled with ProCAST and the results were verified with thermocoupled mold experiments. G and R values were computed from the thermal histories of certain points within the casting.

Upon comparing the actual microstructures to published solidification map predictions, discrepancies were found. Specifically, the criterion function used to predict the formation of freckle defects (critical value of GxR) proved to be incorrect for the castings under study. Consequently, the researchers attempted to find a more accurate criterion function to predict freckle defects and amount of porosity. One criterion function studied was the "Gradient Acceleration Parameter," GAP=(GxR)/t_s, in which t_s is the local solidification time. A function containing local solidification time as well as G and R may be more sensitive to single crystal investment casting. However, this function did not correlate well with observed freckle defects.

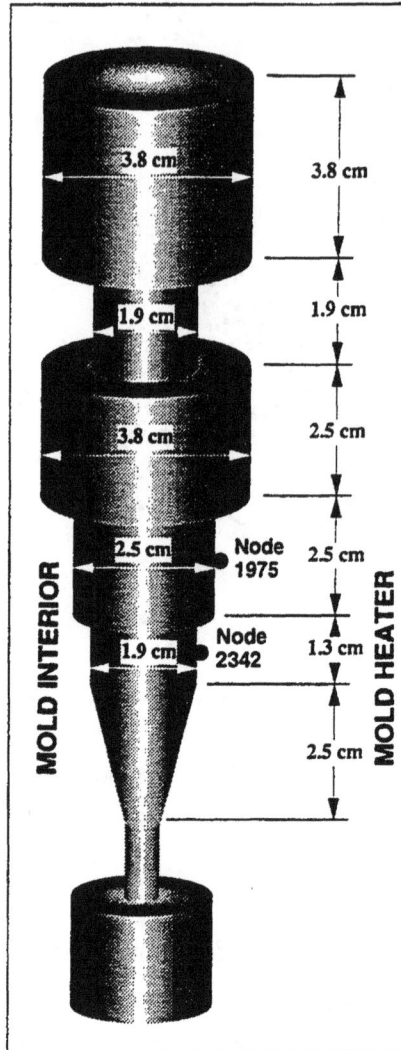

Figure 30. The test casting used by to investigate defects in single-crystal investment castings.

Purvis, et al. also believed that freckle defect formation and porosity might be affected in a similar fashion by changes in solidification conditions. This led them to the Xue porosity function, $XUE = G_s/T_s^{0.5}$. The Xue porosity function correlates better to actual freckle defects than the GAP or the cooling rate criterion (lower values of XUE mean a greater tendency for freckle defects), but a critical value for freckle formation was not found. The last function that was studied was the "directional growth ratio," or

52

the ratio of the solidus isotherm velocity in the withdrawal direction to the growth velocity in the lateral plane. This ratio is sensitive to nonuniformities in heat extraction which can cause a breakdown of directional solidification. This function changed with changes in cross-section, but again no critical value was found for the prevention of freckle defects.

Based on these results, Purvis, et al. concluded that more work is needed in the area of predicting freckle defects in complex shape single crystal investment castings. Specifically, more sensitive criterion functions are needed to describe changes in the solidification front that lead to defect formation, and a clearer understanding of the mechanisms of defect formation is required for accurate microstructure modeling.

3.3. Analytical Modeling of Microstructure Evolution During Solidification

3.3.1. Modeling of Solute Distribution

During alloy solidification, variations in the composition of the solid and liquid occur with time and temperature. Compositional variations during solidification can be modeled analytically for special types of solidification such as normal solidification, in which an entire charge is melted and solidified from one end with a plane front. There are several existing models which differ based on assumptions regarding diffusion and convection. Treatment of many of these models can be found in Flemings[33] and Kurz and Fisher.[23] Several such models are summarized here.

If solidification were to proceed extremely slowly, diffusion would eliminate any segregation and equilibrium solidification would occur. In equilibrium solidification, the solid would be of one composition and the liquid of another. The composition and relative amounts of solid and liquid would change gradually with a change in temperature, as dictated by the equilibrium phase diagram. For a binary alloy, the equilibrium lever rule, $C_s f_s + C_L f_L = C_o$, in which C_s is the composition of the solid, f_s is the fraction solid, C_L is the liquid composition, f_L is the fraction liquid ($f_L = 1 - f_s$), and C_o is the initial liquid composition, determines the composition of the solid and the liquid at any point during solidification (Figure 31[33]). Equilibrium solidification assumes complete diffusion in both the solid and the liquid, which in practice is never observed.

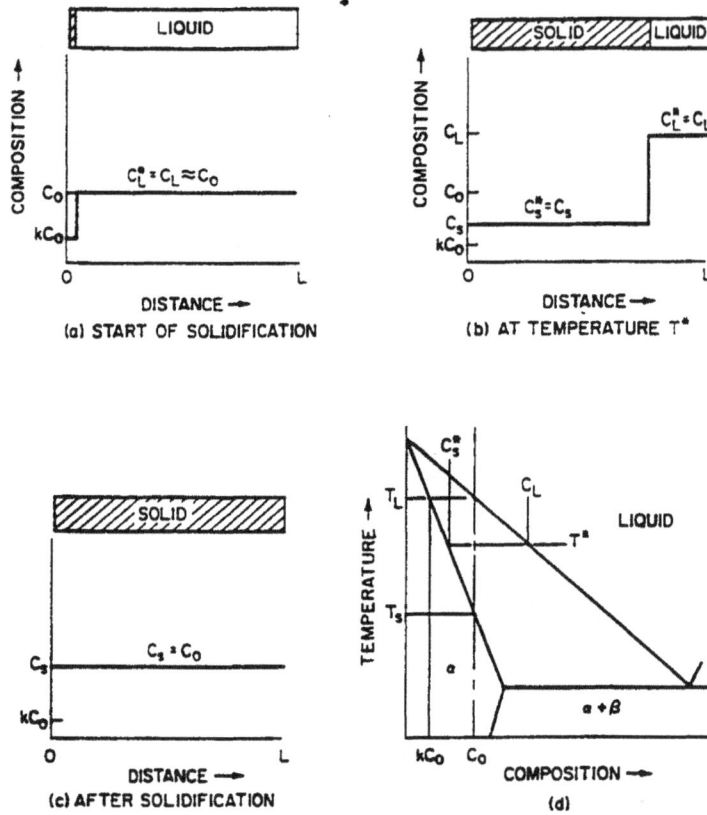

Figure 31. Solute redistribution in equilibrium solidification of an alloy of composition C_o (a) at the start of solidification, (b) at temperature T^*, and (c) after solidification; (d) the corresponding phase diagram.

In attempts to better model solute distribution during solidification, researchers have formulated new problem statements with different assumptions. One particularly useful approach is attributed to Gulliver, Scheil, and Pfann; the most important result of this approach is the "Scheil equation" or the "nonequilibrium lever rule." This model assumes complete diffusion in the liquid, but no diffusion in the solid. Under this model, the solid that forms at each temperature remains at its original composition throughout, while the liquid composition follows the liquidus. The amount of solute rejected upon the freezing of an infinitesimal amount of solid is equal to the amount of solute increase in the liquid.

The equation relating the composition of the solid to the composition of the liquid that results for a binary alloy is $(C_L - C_s^*)df_s = (1 - f_s)dC_L$, in which C_L is the composition of the liquid, C_s^* is the composition of the solid, and f_s is the fraction solid. From this equation, relationships between the composition of the solid at the liquid-solid interface (or of the liquid) and the fraction of solid (or liquid) can be developed. Through integration, it is found that $C_s^* = kC_o(1-f_s)^{(k-1)}$ and $C_L = C_of_L^{(k-1)}$, in which k is the equilibrium partition ratio C_s^*/C_L^*. These equations can be accurate for certain cases of normal solidification. By solving these equations for a given fraction solid, a plot of composition versus distance from the heat sink can be created (Figure 32[33]).

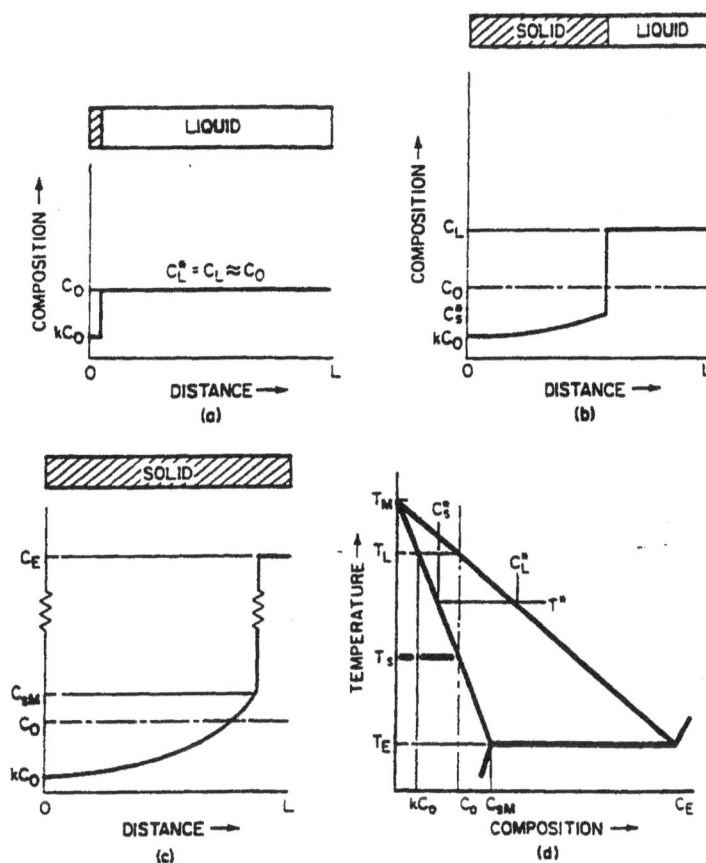

Figure 32. Solute redistribution in solidification with no solid diffusion and complete diffusion in the liquid (a) at the start of solidification, (b) at temperature T^*, and (c) after solidification; (d) the corresponding phase diagram.

More complex models consider the effect of limited liquid diffusion and/or convection. All of these models assume normal, plane front solidification, equilibrium at the liquid-solid interface, and no supercooling of the melt. In reality, these are poor assumptions, but many of these methods can be adapted for use under more realistic conditions by applying them individually to small volume elements.

55

3.3.2. Modeling of Macrostructure

The grain morphology of a casting is often considered to be the macrostructure of the casting. Properties such as macroporosity and macrosegregation can also be considered to be part of the casting's macrostructure. The grain morphology and the existence of macroporosity and macrosegregation can often be predicted analytically.

As was discussed briefly in the solidification map section of this report, castings can exhibit several different morphologies, including equiaxed dendritic, columnar dendritic, cellular, and planar. Most commercial castings contain one or more of three distinct morphological regions - the chill zone, the columnar zone, and the equiaxed zone. The chill zone, or outer equiaxed zone, is typically a small thickness of fine, equiaxed grains which form on the surface of castings due to the initial rapid heat removal at the mold-metal interface immediately after pouring. The columnar zone is a section of columnar dendrites formed by preferential growth of certain equiaxed dendrites from the chill zone. The columnar zone stretches from the end of the chill zone into the center of the casting, where it is followed by a section of equiaxed grains in the inner equiaxed zone (Figure 33[23]).

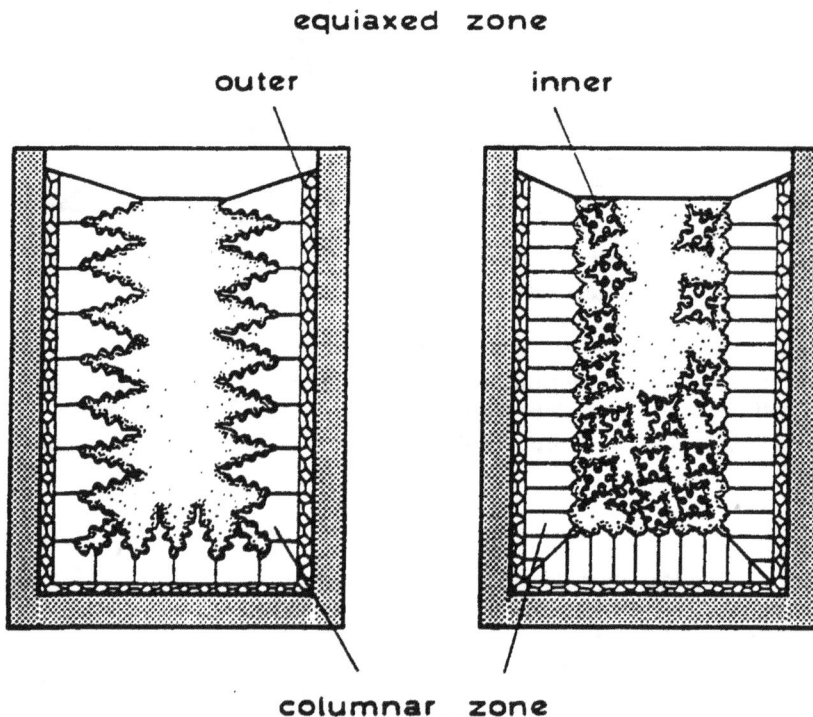

Figure 33. Sketch of the formation of a typical ingot structure showing the chill zone, the columnar zone, and the equiaxed zone.

56

The existence of different solidified morphologies is largely affected by global solidification parameters such as thermal gradient and solidification velocity (Figure 34[23]). As was shown in the solidification map section, critical functions for the transition from one morphology to another can be found empirically. For example, a typical stability criterion for a planar growth morphology is $G/R > (T_L-T_S)/D_L$, in which T_L is the liquidus temperature, T_S is the solidus temperature, and D_L is the diffusion coefficient in the liquid.

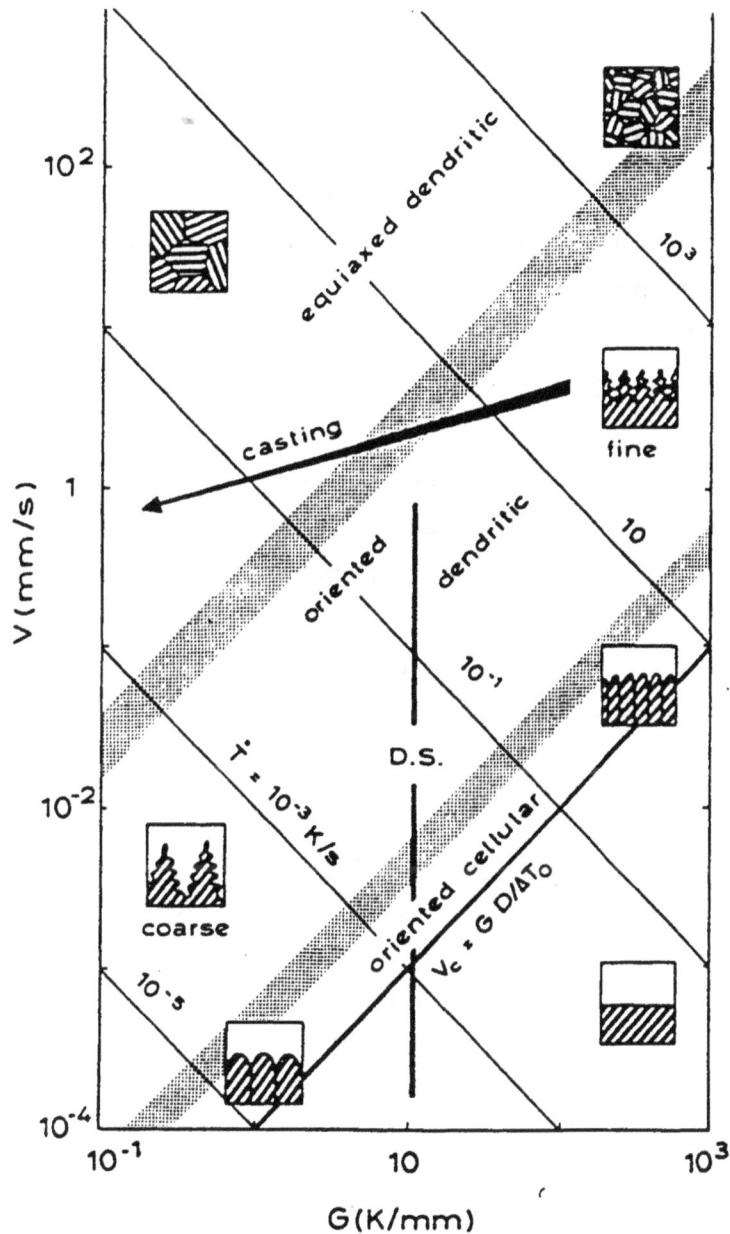

Figure 34. Schematic summary of single-phase solidification morphologies.

3.3.3. Modeling of Solidification Kinetics

The modeling of solidification described above does not address the important topic of solidification kinetics (i.e., nucleation and growth kinetics). Modeling of solidification kinetics can be used to predict cast grain size, secondary dendrite arm spacing, and solidification path. The models described above do not have the ability to predict any of these features of castings without the use of empirical relationships, which are not universally applicable. Overfelt gives a summary of several ways to model nucleation and growth kinetics in casting, and his presentation is summarized here.[18]

A typical nucleation rate for a solidifying alloy is given by the following Dirac delta function: $dN/dT = N_s \cdot \delta(T-T_N)$, in which N is the number of nuclei, N_s is the number of active substrates, T is the temperature, and T_N is the critical nucleation temperature for the active substrate. The combined effect of all active substrates can be found through superposition of all relevant nucleation rate equations. Experimental data for the number of active substrates can be fit to the equation $N_s = K_3 + K_4(dT/dt)^2$, in which K_3, K_4, and T_N can be found by conducting experiments at various cooling rates.

For growth of a nucleated grain, growth rate is given by $R = \mu(\Delta T)^2$, in which R is the growth rate, μ is an alloy-dependent constant, and ΔT is the undercooling. For equiaxed grains, grain size is different from fraction solid, which is given by $\Delta f_s(t) = n(t) \cdot [4\pi r^2(t) \cdot \Delta r(t) \cdot f_i(t) + (4/3)\pi r^3(t) \cdot \Delta f_i(t)]$, in which n(t) is grain density, r(t) is the radius of the spherical envelope of the equiaxed grain, and $f_i(t)$ is the percent of internal solid fraction.

Other relationships for nucleation and growth kinetics exist in the literature.[33],[23] Most involve parameters that must be determined experimentally. Once nucleation and growth laws are known for a given alloy, they can be incorporated into heat, fluid, and solute flow models to obtain advanced solidification models. These advanced solidification models can then be used to predict as-cast grain size and other microstructural features.

3.4. Observation and Interpretation of Cast Titanium Alloy Microstructures

No solidification maps currently exist for titanium alloys. As the demand for titanium castings increases, the ability to predict as-cast microstructure accurately will become important. A prerequisite for the development of these predictive capabilities is the understanding of the important features of cast titanium microstructures. This requirement guides the present discussion.

3.4.1. General Titanium Microstructure

The microstructure of a titanium alloy component is highly dependent upon the alloy composition and the thermomechanical history of the component. Titanium alloys can have many different crystal structures, including alpha (HCP), beta (BCC), alpha-2 (Ti_3Al - DO_{19}), gamma (TiAl - $L1_0$), or some combination of these and various metastable phases such as alpha prime (hexagonal martensite), alpha double prime (orthorhombic martensite), or omega, depending on composition and temperature. The microstructure is highly dependent on composition, heat treatment, working history, etc. It can be equiaxed, lamellar, or a mixture of the two (duplex). Some example titanium microstructures for several different alloys are shown in Figure 35.[3]

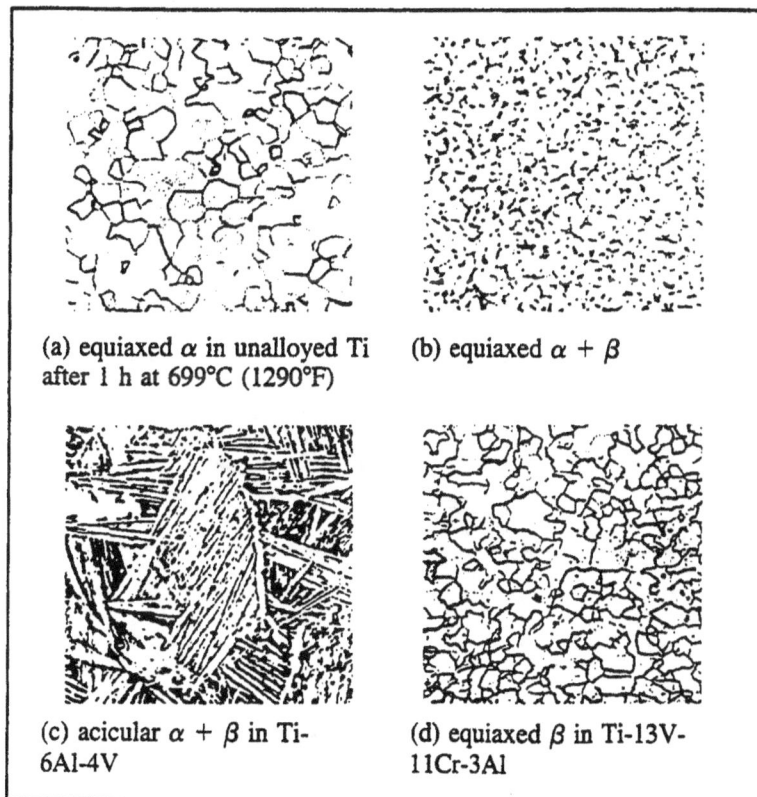

(a) equiaxed α in unalloyed Ti after 1 h at 699°C (1290°F)

(b) equiaxed $\alpha + \beta$

(c) acicular $\alpha + \beta$ in Ti-6Al-4V

(d) equiaxed β in Ti-13V-11Cr-3Al

Figure 35. Typical microstructures of α, $\alpha + \beta$, and β-Ti alloys.

59

The evolution of microstructure of Ti alloys during solidification has not been deeply explored. Some research has shown that it can be difficult to deduce the evolution of microstructure of Ti alloys using room temperature metallography techniques because the post-solidification transformation to the Widmanstätten structure masks the original structure and because large amounts of solid state diffusion take place in Ti at elevated temperatures.[34] Little is known about the critical values of parameters such as thermal gradient G and growth velocity R for predicting the transition from a columnar to equiaxed dendritic growth morphology or the types and distribution of casting defects.

Some work has been done on as-cast morphology and segregation in certain binary Ti alloy systems.[35] This work attempted to relate dendrite arm spacing, local solidification time, composition, and degree of microsegregation to one another. Empirical relationships between local solidification rates and secondary dendrite arm spacings of the form $d = C\theta_f^n$, in which d is the secondary dendrite arm spacing, θ is the local solidification time, and C and n are empirically determined constants, were developed (Figure 36[35]). Measured microsegregation ratios were measured and plotted against weight percent solute (Figure 37[35]). General conclusions were that (1) microsegregation at the end of freezing and at room temperature can be quite different, (2) local solidification time has the greatest effect on dendrite arm spacing, and (3) for similar cooling rates, secondary dendrite arm spacing decreases with increasing solute content for most alloy systems.

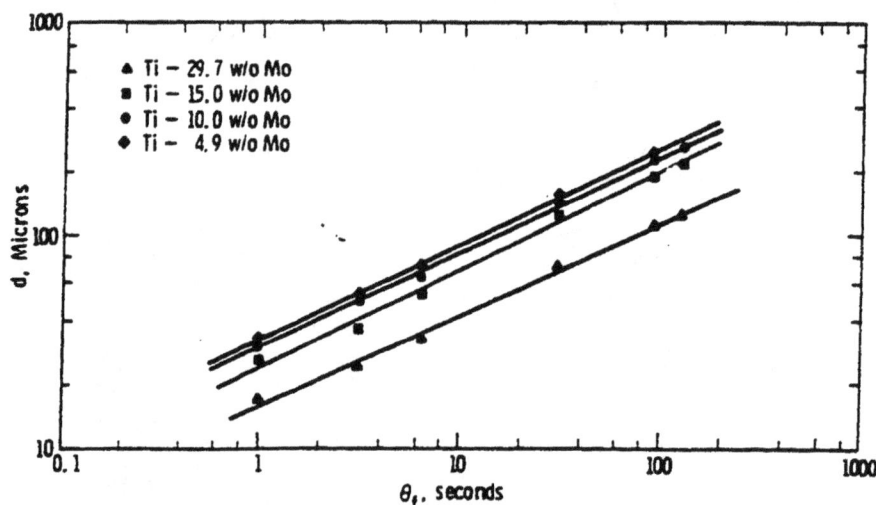

Figure 36. Plot of secondary dendrite arm spacing versus local solidification time for four Ti-Mo alloys.

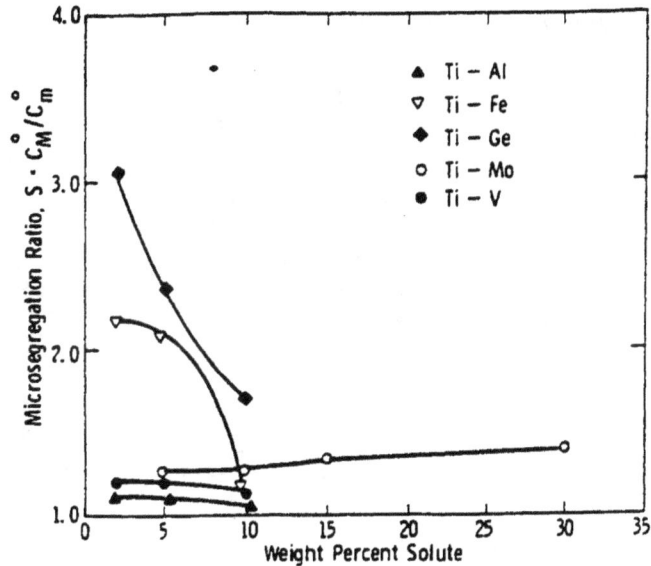

Figure 37. Segregation ratio versus local solute content measured for alloys from five titanium base alloy systems. Mesurements were on samples of similar dendrite arm spacing.

3.4.2. Cast Ti-6Al-4V Microstructure

The popular Ti-6Al-4V alloy is an alpha + beta alloy. Ti-6Al-4V solidifies as 100% beta with morphology and grain size dependent on thermal conditions. Typical Ti-6Al-4V castings solidify as dendritic beta with grain sizes ranging from 0.5 to 5 mm.[6] As the beta cools into the alpha + beta phase field, alpha formation commences along the beta grain boundaries. Upon further cooling, some of the beta begins to transform to alpha platelets on specific crystallographic planes of the original beta by a burgers relationship, and an acicular or Widmanstätten structure develops. Slow cooling results in the formation of alpha platelet colonies (i.e., colonies of alpha platelets with similar alignment and crystallographic orientation). Rapid quenching from above the beta transus results in a martensitic transformation that produces a fine Widmanstätten structure.

Cast titanium alloys such as Ti-6Al-4V can be HIP'd and heat treated (stress relieved, process annealed, or solution treated and aged) to heal internal porosity and to alter the microstructure to promote certain properties such as machinability, hardness, or ductility. Proper heat treatment can be used to eliminate large alpha platelet colonies and grain boundary alpha, resulting in improved tensile and fatigue properties. The final cast + HIP'd microstructure of Ti-6Al-4V is very similar to the microstructure of a beta-processed wrought Ti-6Al-4V microstructure (Figures 38[36] & 39[6]).

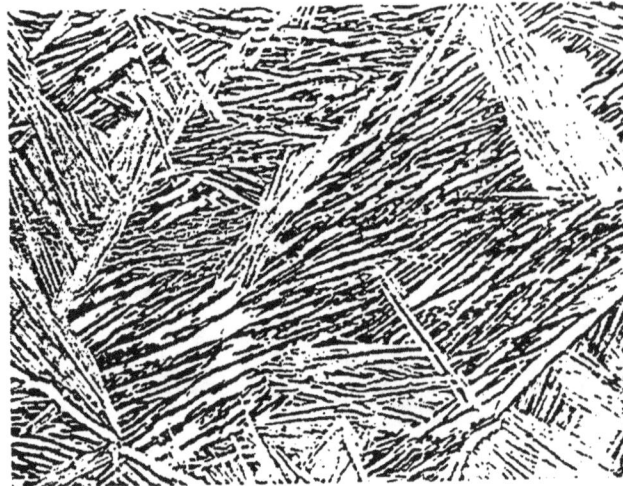

Figure 38. Sample microstructure of a titanium casting; 200X.

Figure 39. Typical microstructure of a cast Ti-6Al-4V component showing mixed α + β colony structure; 200X.

3.4.3. Cast Gamma TiAl Microstructures

McCullough et al.[34] conducted research on solidification of gamma titanium aluminide alloys. Their work focused mainly on determination of the Ti - Al phase diagram, but contained considerable information regarding microstructural evolution during solidification. The primary solidification phases were determined by inspection of the dendritic structure in shrinkage cavities of arc melted buttons. Solid state phase transformations were explored for compositions between 40 and 55 at. % Al. Some of their conclusions included the following: (1) regardless of the primary phase to solidify, the microstructures of Ti- 40-49% Al alloys are nearly indistinguishable from one another; the final microstructure is equiaxed colonies of the [alpha-2 + gamma] lath structure, and no dendrite pattern is distinguishable, (2) in alloys with aluminum content greater than 45%, gamma segregate forms between the grains of equiaxed [alpha-2 + gamma] laths, (3) in alloys of composition Ti- 48-52% Al, cellular regions consisting of alpha-2 and gamma form between the laths and the gamma segregate; the cellular gamma is related to the gamma segregate, and the cellular alpha-2 is related to the alpha-2 in the lath structure. The evolution of the microstructures of several such alloys is described below. Figure 40[34] shows some cast gamma TiAl microstructures and Figure 41[34] shows the Ti-Al phase diagram for reference.

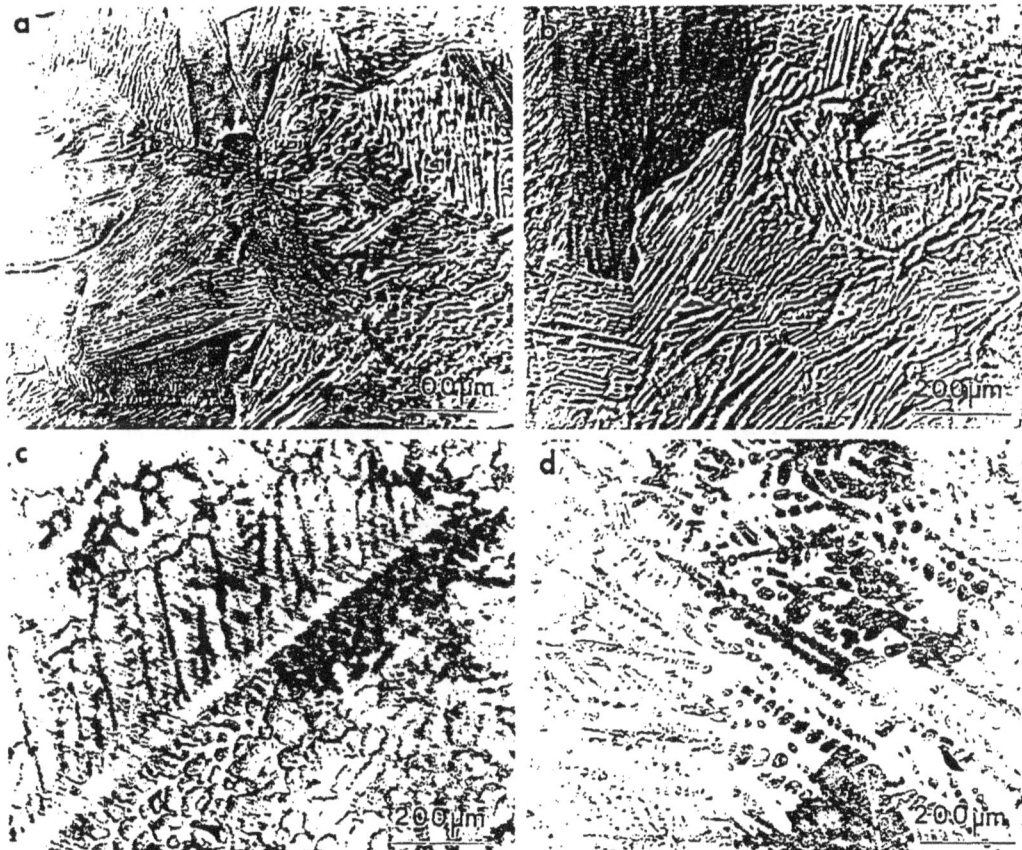

Figure 40. Representative γ-TiAl microstructures: (a) Ti-40 at.%Al, (b) Ti-45 at.% Al, (c) Ti-50 at.% Al, and (d) Ti-55 at.% Al

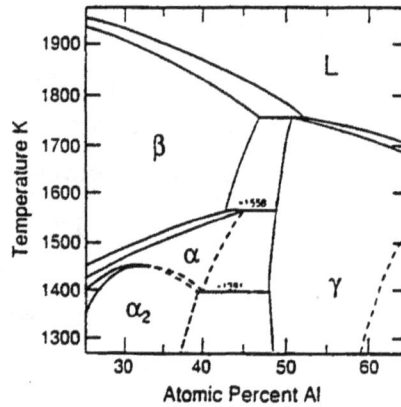

Figure 41. The Ti-Al binary equilibrium phase diagram.

Ti- 40-49% Al: As the molten alloy is cooled from above the liquidus, the first phase to solidify is dendritic beta. When the peritectic temperature is reached, the remaining liquid transforms to alpha, which surrounds the beta dendrites. As further cooling occurs, the beta becomes unstable and transforms to 100% alpha. This transformation proceeds according to the burgers orientation relationship, and the basal plane of the alpha will lie on one of the six {110} planes of the beta. With additional cooling, the alpha becomes unstable and transforms to alpha-2 by an ordering transformation. This ordering transformation results in the formation of stacking faults and anti-phase boundaries, upon which plates of gamma can nucleate and grow. The nucleation and growth of gamma plates on planes parallel to the basal plane of the alpha-2 results in a final microstructure consisting of large equiaxed colonies of [alpha-2 + gamma] laths with six possible orientations within a prior beta dendrite. For compositions above 45% Al, gamma segregate is formed in the interdendritic spaces.

Ti- 50-55% Al: For these alloys, the primary solidification phase is dendritic alpha. Upon cooling, these alpha dendrites are surrounded by gamma segregate through a peritectic reaction. The alpha transforms to alpha-2 through an ordering transformation, and the gamma plates form on a single {111} habit plane parallel to the basal plane of the alpha-2 dendrites. The resulting structure is again equiaxed colonies of [alpha-2 + gamma] laths with gamma segregate, but in this case, all of the laths within a single primary dendrite have the same orientation.

3.5. Microstructure Observations in Titanium Powders

A large amount of work has been done in the area of titanium powder metallurgy because of the unique structures and correspondingly unique mechanical properties that can be obtained. Because many titanium powders are formed by so-called "rapid solidification" techniques, a description of the microstructures found in titanium powders is appropriate for this report.

Titanium alloy powder microstructures depend mainly on alloy composition and cooling rate both during and after solidification. Rapidly solidified titanium powders have increased solubility limits for most alloying elements. This increased solubility results in a modification of the expected microstructure for specific alloys. Rapid solidification can produce microstructures that aren't normally seen in titanium alloys, such as finer grained, more homogeneous conventional structures and even previously undiscovered metastable phases such as equiaxed alpha in very fine grained Ti-6Al-4V and fine dispersoids in supersaturated solid solutions.[37] Some representative titanium powder microstructures are shown in Figures 42[38], 43[39], 44[40], 45[41] and 46.[42]

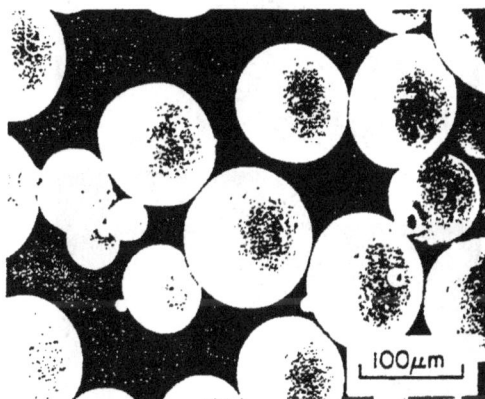

Figure 42. Gas atomized Ti-6Al-4V powder.

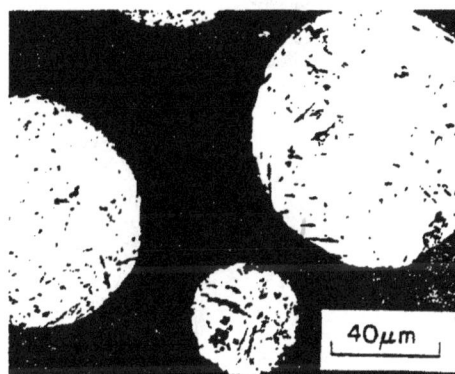

Figure 43. Microstructure of gas atomized Ti-6Al-4V powder.

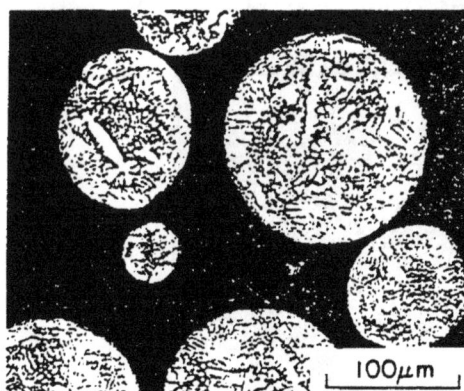

Figure 44. Microstructure of PREP Ti$_3$Al base powder; 250X.

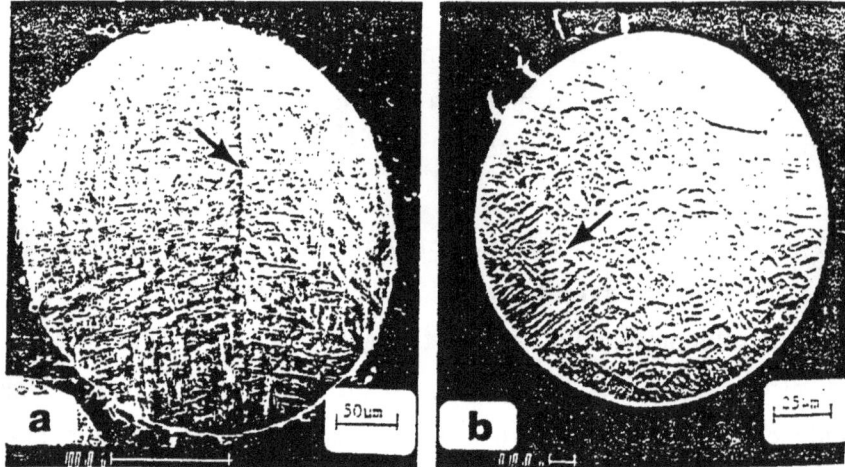

Figure 45. SEM images showing microstructures of (a) REP Ti-6Al-4V powder particles; 250X, and (b) PREP Ti-10V-2Fe-3Al powder particles; 500X. (Arrows mark beta grain boundaries.)

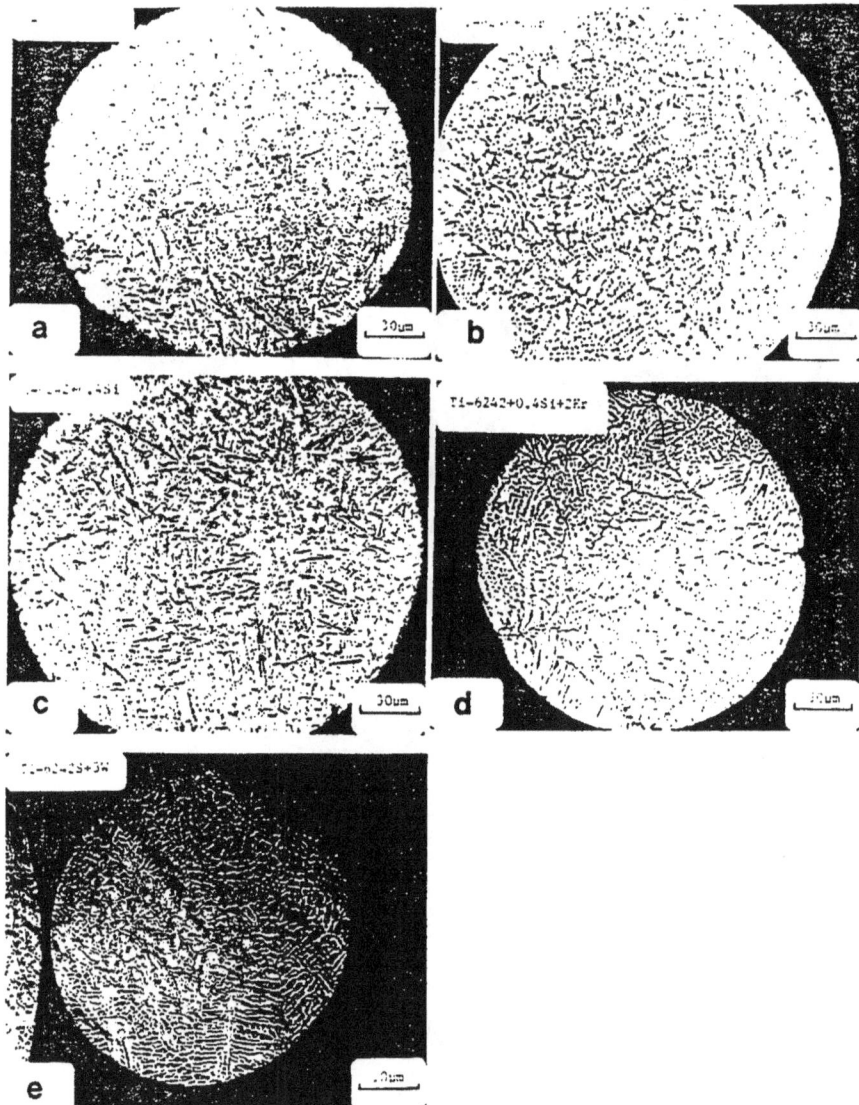

Figure 46. As-solidified PREP powder particles: (a) Ti-6242 + 2Er, (b) Ti-6246 + 2Er, (c) Ti-6242 + 0.4Si, (d) Ti-6242 + 0.4Si + 2Er, and (e) Ti-6242 + 3W.

Rapidly solidified Ti-6Al-4V can have several structures. Upon cooling from the beta phase field, the beta undergoes a martensitic transformation in the same way as conventionally solidified titanium, but the microstructure is finer grained. If the cooling rate is slow enough to produce a beta grains size above ~5 microns, normal lenticular alpha forms upon reheating. If the rapidly solidified beta grain size is below ~5 microns, the formation of alpha from martensite occurs via a different mechanism. The alpha grows from the martensite in an equiaxed structure instead of a lenticular structure (Figure 47[42]). This difference in morphology is believed to be caused by the dominance of grain boundary alpha due to the small diffusion distances in the fine grained material.[42]

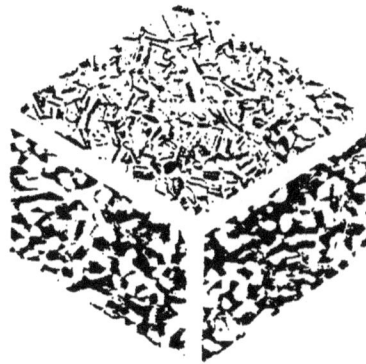

Figure 47. Equiaxed alpha morphology in Ti-6Al-4V produced after annealing fine grained rapidly solidified material.

An interesting result of titanium powder metallurgy research is that when the beta grain size of Ti-6Al-4V powder was plotted against the estimated cooling rate during powder formation on a log-log scale, a straight line with a slope of -0.57 resulted regardless of powder production process(Figure 48[42]). This result could be useful in the study of conventional casting processes.

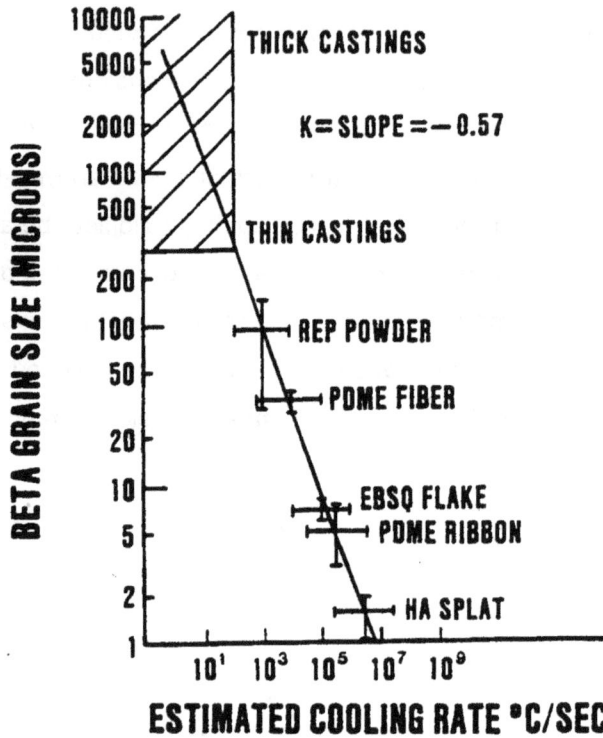

Figure 48. Effect of cooling rates on prior beta grain size of Ti-6Al-4V alloy.

3.6. Discussion of Research Issues/Approaches

Research issues in modeling of microstructure evolution in PMC of titanium alloys include:

- The construction and validation of solidification maps for specific titanium alloys
- The development of useful relationships between casting conditions (cooling rate, solidification velocity, melt superheat, initial composition) and solute distribution on both a microscopic and macroscopic scale
- The development of useful nucleation and growth kinetics relationships for specific titanium alloys

Research on titanium alloy microstructure evolution during solidification and subsequent cooling is needed before manufacturers can accurately predict as-cast microstructures. An approach similar to the approaches described above could be used to generate solidification maps for various titanium alloys. Specifically, directional solidification experiments could be conducted and various criterion functions could be

68

evaluated for use with Ti-6Al-4V and a gamma TiAl alloy. Relationships between grain shape/size and cooling rate, etc., could be identified and evaluated with directional solidification experiments. The results could then be validated with actual permanent mold casting trials and modeling. The coupling of solidification conditions to solute distribution and cast macrostructure could be achieved using a similar approach. Due to the great amount of solid-state diffusion that occurs in titanium alloys at high temperatures, solid-state cooling cannot be ignored in these studies; hence, the incorporation of post-solidification beta grain growth into any grain size-cooling rate relationship would be interesting.

4.   Modeling of Solidification

Modeling of solidification processes requires an understanding of heat, fluid, and mass transfer. Solidification modeling can be approached from the atomic level (nucleus formation and atomic attachment kinetics), the microscopic level (dendritic growth and microsegregation), or the macroscopic level (macroscopic heat and fluid flow). Coupling these approaches to solidification into one all-encompassing model of solidification is not a simple matter. Typically, each aspect of solidification is modeled separately, and macroscopic properties are used to make qualitative inferences about microstructure. With increasing computing speeds, complex models incorporating both micro- and macro-scale phenomena can now be used to predict microstructural evolution in certain cases.

Modeling of the casting process can be extremely useful for designing molds and process parameters. Casting defects, cycle time, and mold wear can be minimized and casting yield can be maximized. The use of solidification modeling for designing molds (placement and size of gates, risers, chills, etc.) can help eliminate or minimize defects such as porosity in castings. Modeling of the mold filling process can be used in mold/gating design to ensure complete mold fill and to minimize mold wear by erosion and abrasion. In addition, modeling can be used to determine operating windows or to predict optimum process parameters such as superheat, mold preheat or mold cooling, and dwell time in the mold. Modeling can also be used to predict thermal stresses in castings and molds, which in turn can be used to predict hot tearing or heat checking. Figure 49[18] shows some of the areas of casting that can be modeled currently.

There are several common problems encountered when modeling casting processes. A major problem is the difficulty in finding accurate thermophysical data which are required input for solidification modeling. Interface heat transfer coefficients (IHTCs) are difficult to determine and are system dependent; hence, finding the proper values for the system to be modeled is not trivial. In addition, incorporating the time and/or temperature dependence of the thermal conductivity, specific heat, density, or IHTC complicates the model. For alloy solidification, measurement and proper distribution of the latent heat of fusion over the freezing range can be a problem. Also, because so many solidification characteristics are alloy dependent, it is nearly impossible to form a global solidification modeling technique.

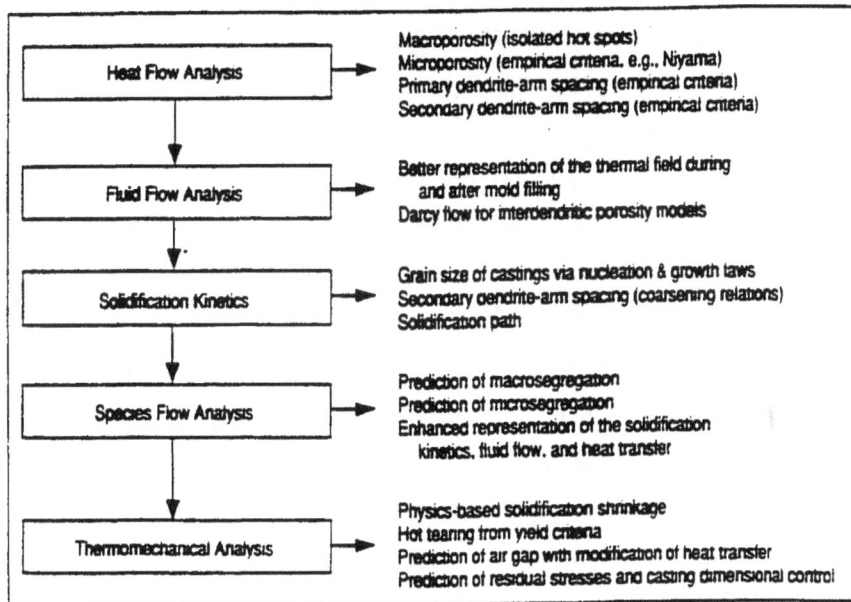

Figure 49. The types of analyses available for solidification modeling and their benefits.

This chapter reviews some analytical and numerical approaches to modeling solidification processes. Some examples of the use of modeling in casting from the literature are reviewed. Special attention is given to related issues such as determination of thermophysical data and interface heat transfer coefficients. Future research issues and approaches are addressed in the final section.

4.1. Analytical Modeling Approaches

Analytical solutions to the solidification problem are only available for simple shapes with simple boundary conditions or for cases in which many simplifying assumptions are invoked. Macroscopically, solidification depends on heat diffusion and convective fluid flow, and the governing differential equations for these transport mechanisms can be solved analytically for special geometries and boundary conditions. Presentations of various forms of solidification problem solutions can be found in Flemings[33], Kurz and Fisher[23], Gaskell[43], and Carslaw and Jaeger.[44] Some solidification problems in which analytical solutions are possible are discussed below.

Analytical solutions to the heat transport problem in solidification exist for a pure metal initially at its melting point solidifying under conditions of one dimensional heat flow. The differential equation for one dimensional heat transport is $\partial T/\partial t = \alpha \partial^2 T/\partial x^2$ in which T is temperature, t is time, x is distance, and α is thermal diffusivity. This equation can be solved analytically if proper initial and boundary conditions prevail.

As an initial example, consider a pure metal at its melting point solidifying in a flat, insulating mold (Figure 50[33]). The initial conditions are the mold is at room temperature (T_o), and the liquid is at its melting point (T_m) at t=0. The boundary conditions are $T=T_m$ at the mold-metal interface and within the solidifying metal (for $x \geq 0$), and $T=T_o$ at the outer mold wall. For an insulating mold, it is assumed that there is no thermal gradient in the solidifying liquid, which remains at its melting point throughout, and the thermal gradient in the mold increases from room temperature on the outside of the mold to the melting temperature of the solidifying material on the inside.

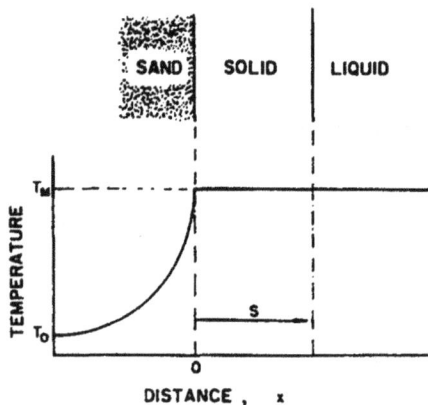

Figure 50. Approximate temperature profile in solidification of a pure metal poured at its melting point against a flat, smooth mold wall.

The solution to the differential equation in the mold for the stated boundary conditions is $\{(T-T_m)/(T_o-T_m)\} = \text{erf}\{-x/[2\sqrt{(\alpha_m t)}]\}$, which gives a temperature distribution in the mold as a function of time and position. To obtain a relation between time and thickness solidified, S, equations for conductive heat flow must be used to form a heat balance across the mold-metal interface. The rate of heat flow into the mold at the mold-metal interface is given by $(q/A)_{x=0}=-K_m(\partial T/\partial x)_{x=0}$, in which K_m is the thermal conductivity of the mold. Differentiating the solution to the differential equation with respect to x, setting x=0, and combining with the equation for the rate of heat flow into

72

the mold results in $(q/A)_{x=0} = -\sqrt{[(K_m\rho_m c_m)/\pi t]}(T_m-T_o)$, in which ρ_m and c_m are the density and specific heat of the mold, respectively. Recalling that the only heat input in this problem is due to the heat of fusion of the solidifying material gives $(q/A)_{x=0} = -\rho_s H(\partial S/\partial t)$, in which H is the latent heat of fusion of the solidifying material. Combining the two heat flow equations gives $-\rho_s H(\partial S/\partial t) = -\sqrt{[(K_m\rho_m c_m)/\pi t]}(T_m-T_o)$, which can be separated and integrated with respect to time and thickness solidified from t=0 and S=0 to give the final result $S = (2/\sqrt{\pi})\{(T_m-T_o)/\rho_s H\}\sqrt{(K_m\rho_m c_m)}\sqrt{t}$, or a square root of time dependence for thickness solidified. The interfacial condition that gives a square root of time dependence of thickness solidified is called the "ideal interface" condition.

This solution is strictly valid only for a metal of high thermal conductivity solidifying within a flat, highly insulated mold, i.e., in a case in which the assumed boundary and initial conditions are most valid. Hence, the solution has limited applicability for more complex situations. In the case of a mold that is of a more complex shape, an approximate solution can be obtained by replacing solidified thickness (S) by the quotient of the volume solidified by the surface area of the mold-metal interface (V_s/A). By this means, the time required for the entire casting to freeze can be predicted. This solution, $t_f = C(V/A)^2$, in which C is a constant, is known as Chvorinov's rule. Similar solutions can be found for other simple shapes such as spheres and cylinders. Casting problems for which this solution might be valid are sand casting and investment casting. Casters often use Chvorinov's rule in practice to get a rough idea of the time required for a particular complex casting to solidify without the use of more time consuming modeling methods.

When the resistance of the mold-metal interface dominates heat flow, a different analytical solution exists for the one-dimensional, flat mold problem. In this case, it is assumed that the mold remains at room temperature and the solidifying melt remains at its melting point (Figure 51[33]). The only change in temperature occurs at the metal-mold interface. Again, the heat flow into the mold is given by $(q/A)_{x=0} = -\rho_s H(\partial S/\partial t)$. Because the interfacial resistance to heat flow dominates, $(q/A)_{x=0} = -h(T_m-T_o)$, in which h is the interface heat transfer coefficient. Combining and separating these two equations and then integrating with respect to t and S from t=0 and S=0 results in $S = h[(T_m-T_o)/\rho_s H]t$, or a linear time dependence of thickness solidified. This linear time dependence occurs when the interface is a "Newtonian interface".

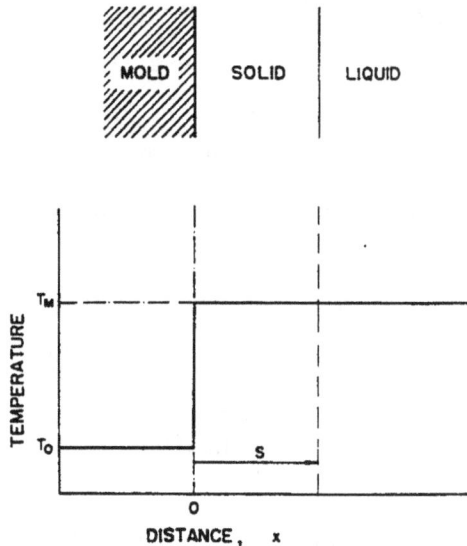

Figure 51. Temperature profile during solidification against a large flat mold wall with mold-metal interface resistance controlling.

The interface resistance solution is most valid for a highly conductive metal freezing in a highly conductive mold. In this case, shape does not affect the heat transfer across the interface, so S can be replaced by V/A and solutions for more complex shapes can be found without any loss of accuracy. Permanent mold casting, die casting, and freezing against a chill are problems for which this solution might be appropriate. Two criteria that can be used to decide if the heat transfer across the interface dominates (i.e., if the above solution is valid) are $h \ll K_s/S$ for conductive molds and $h^2 \ll K_m \rho_m c_m/t$ for insulating molds.

Both solutions discussed above have severe limitations. Very few situations truly approach either of these two (ideal or Newtonian) cases. Most often solidification is affected by a combination of mold/metal heat conduction properties and interface heat transfer properties, or is "mixed mode" in nature. The boundary and initial conditions are more complex, and correspondingly, solutions are more complex.

Analytic solutions do exist for the case in which interface resistance is small and the mold is held at a constant temperature (i.e., a water-cooled chill) or the mold is very thick. For example, consider a pure material freezing against a flat, water-cooled chill (Figure 52a[33]). The initial conditions are the temperature of the mold is given ($T=T_o$ for $x \leq 0$ and t=0), and the temperature of the solidifying material is the melting point ($T=T_m$ for $x > 0$ at t=0). The boundary conditions for this case are (1) at the mold

74

wall, the temperature is known and constant ($T=T_0$ at x=0), and (2) at the solid-liquid interface, the temperature is the melting temperature of the material ($T=T_m$ at x=S) and $K_s(\partial T/\partial x)_{x=S} = H\rho_s \partial S/\partial t$, in which K_s is the thermal conductivity of the solidifying material, r_s is the density of the material, H is the heat of fusion, and S is the thickness solidified. The temperature distribution within the solidifying material is given by $(T-T_0)/(T_0{'}-T_0) = \text{erf}\{x/[2\sqrt{(\alpha_s t)}]\}$, in which $T_0{'}$ is an integration constant. Conducting a heat balance across the mold-metal interface and differentiating the temperature distribution with respect to x to obtain an expression for $\partial T/\partial x$ in the solidifying material gives an expression which can be integrated to find the thickness solidified versus time. The result is $S = 2g\sqrt{(\alpha_s t)}$, in which g is determined from $ge^{g^2}\text{erf } g = (T_m-T_0)C_s/(H\sqrt{\pi})$.

Changing from a chill to a semi-infinite mold changes the problem and the results slightly (Figure 52b[33]). The temperature of the mold is now allowed to vary, but the outside mold wall temperature and the mold-metal interface temperature remain constant. g is given by $ge^{g^2}(\sqrt{[(K_s\rho_s C_s)/(K_m\rho_m C_m)]})\text{erf } g = (T_m-T_0)C_s/(H\sqrt{\pi})$ and the temperature distribution is given by $(T-T_s)/(T_0{'}-T_s) = \text{erf}\{x/[2\sqrt{(\alpha_s t)}]\}$ in which T_s is the temperature of the mold-metal interface.

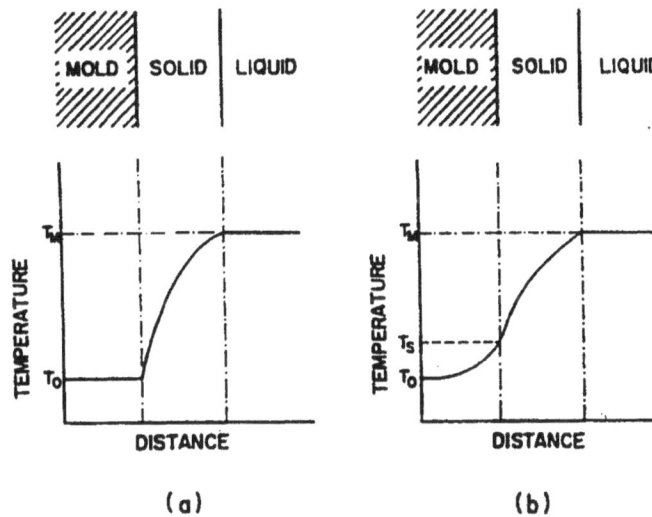

Figure 52. Temperature profile during solidification against a flat mold wall when (a) resistance of the solidifying metal is controlling and when (b) combined resistances of metal and mold are controlling.

75

The solutions discussed so far are all for pure metals. In the case of alloy solidification, there is not a single melting/freezing point. Instead, there is a freezing range, which results in the formation of a "mushy zone" during solidification. At each location in the casting, there is a local freezing time which is the time between the passing of the liquidus and solidus isotherms. Carslaw and Jaeger[44] have presented an analytical solution to an alloy solidification problem whose primary assumptions comprise a semi-infinite metal and mold, no interface resistance, constant thermal properties, and heat of fusion distributed evenly over the solidification range. Other more realistic alloy solidification problems often must be solved using approximate or numerical techniques.

For cases of multi-dimensional heat flow, analytical solutions are rarely available, and numerical methods must be used. Sometimes multi-dimensional problems can be reduced to one-dimensional problems with a few simplifying assumptions, in which case one-dimensional analytic solutions can be used as an approximation.

Experimental agreement with some of these analytic solutions has been good for simple shapes, although the measured time-thickness solidified curves are often displaced to the right on the time axis. This displacement occurs because the molten metal is not poured exactly at its melting point, but instead is poured with a small amount of superheat. The time required to eliminate the superheat shows up in the results as a displacement to the right along the time axis. Agreement is best for cases in which the assumed one-dimensional heat flow and initial and boundary conditions are most valid.

4.2. Numerical Modeling Approaches

Numerical modeling techniques provide solutions to complex, non-linear problems. Typically, they involve the discretization of the governing differential equations of heat and mass transport and the pertinent boundary conditions. The object to be modeled is also discretized into individual volume elements. Both finite element and finite difference methods are currently used to solve solidification problems. Both involve the discretization of the heat conduction equation, $\partial(\rho C_p T)/\partial t = \mathrm{div}\,(k\,\mathrm{grad}\,T) + Q$, in which Q is the volumetric rate of heat generation.

In finite difference methods, space is discretized into a rectangular array. Usually, Taylor series approximations (truncated Taylor series) are used to discretize the differential equations and the boundary and initial conditions to a minimum second order accuracy in both time and space. Alternatively, a control volume discretization can be used along with an assumed point-to-point variation profile. The problem is thereby transformed into a linear algebra problem, and various explicit and implicit discretization techniques and direct or iterative solution techniques can be used to find the value of the dependent variable at each grid point at specific times.

Finite element methods require the discretization of space into finite, non-overlapping elements. The governing equations are discretized using a variational formulation or a form of the method of weighted residuals such as the Galerkin method. The approximate solution consists of the dependent variable values for each node and the between-node interpolation or "shape" functions. Finite element methods are more commonly used for stress modeling, and less commonly used in fluid and heat flow modeling. A finite element method is used in the commercial software package ProCAST, which is discussed in the commercial software section below.

4.2.1. Commercial Software

Solidification modeling packages have many capabilities. These packages often can model fluid flow, heat flow, species flow, and stress generation. Mold fill, solidification, and defect generation can be modeled through the concurrent modeling of various aspects of casting. Through the use of criterion functions or other empirically derived expressions, solidification modeling packages can automatically predict high risk areas for problems such as mold wear, porosity, and hot tearing. There are several commercial software packages designed to model solidification and related phenomena. A review of the methods and capabilities of one such package, ProCAST, is presented here.

ProCAST is a commercial software package from UES, Inc. designed to model industrial castings. It has full three dimensional capabilities, and can be used to solve conjugate heat transfer and fluid flow problems. It can model heat flow in both the casting and the mold, and can handle time and/or temperature dependent properties. The entire casting cycle can be modeled with ProCAST, from filling to mold opening. Turbulent flow, non-Newtonian flow, trapped gas, eddy current heating, solidification kinetics, solid state phase transformations, and elastic, plastic, and elastoviscoplastic

77

stresses can be modeled in certain cases. Local solidification times, macroporosity, and trapped gas porosity can all be predicted.

The use of ProCAST and similar packages to model actual castings in order to predict structure and defects will be addressed in a later section. The focus of this section is on the numerical methods used to solve the differential equations relating to solidification such as the conduction, momentum, and pressure equations.

ProCAST uses a finite element method with space discretized into brick, wedge, or tetrahedral elements. An approximate solution for the given dependent variable is assumed to be a function of the temperatures at the nodes (intersections between elements) and the assumed interpolating functions (which describe the way the dependent variable changes within the elements). The relevant differential equations are discretized in space, approximate solutions are inserted into the equations, and the residual errors are minimized using the Galerkin form of the Method of Weighted Residuals. Time stepping is done using a two-level predictor-corrector numerical integration technique. A description of the discretization and solution of the transient, non-linear conduction energy equation from the ProCAST manual[45] follows. Other differential equations used by ProCAST are handled in a similar manner, and will not be treated here.

ProCAST uses the enthalpy formulation of the conduction equation as it applies to solidification. The resulting energy equation for transient, non-linear conduction is $\rho(\partial H/\partial T)(\partial T/\partial t)$ - div[k grad T] - q(x) = 0 in which enthalpy is given by $H(T) = \int_0^T c_p d\tau + L[1-f_s(T)]$, and L is the latent heat of fusion. In ProCAST, the domain over which the energy equation is applicable is divided into non-overlapping, space-filling elements. The temperatures within these elements are interpolated from temperature values at discrete nodes. An approximate solution of the form $T(x,t) = N_i(x)T_i(t)$ is assumed, in which the N_i's are the (element type dependent) interpolating functions and the T_i's are the nodal temperatures.

Upon inserting the approximate solution into the energy equation, a residual error results. The Galerkin form of the Method of Weighted Residuals is used to minimize the resulting error. This method uses the interpolating functions as the weighting functions in the weighted residuals method. The symmetric matrix system CT' + KT = F results, in which C is the capacitance matrix, $C_{ij} = \int_\Omega \rho(dH/dT)N_iN_jd\Omega$, K is the conductivity matrix, $K_{ij} = \int_\Omega \text{grad } N_i \, (k \text{ grad } N_j) d\Omega + \int_{\Gamma_2} h N_i N_j d\Gamma_2$, and F is the source vector, $F_i = \int_{\Gamma_2} N_i (q - h T_a) d\Gamma_2$.

The integrations for C, K, and F are performed on an element-by-element basis using numerical techniques and the results are assembled to form global C, K, and F matrices. The resulting differential equation, $CT' + KT = F$, is solved using a two-level predictor-corrector numerical integration scheme. In this scheme, a predicted temperature profile is corrected repeatedly until the difference in temperatures between the current and previous iterations is less than a user-specified amount or until a user-specified maximum number of iterations is reached. If the solution does not converge in the specified number of corrector steps, the time step is automatically reduced and the entire process is repeated.

For the spatial discretization described above, the predictor step is $[C + \Delta t \theta K] T_o^{n+1} = [c - \Delta t(1-\theta)K]T^n - \Delta F$, in which C, K, and F are evaluated with temperature values from time n, T_o^{n+1} is the predicted temperature, Δt is the current time step, and θ is a constant that determines the type of discretization used. The corrector step is $[C + \Delta t \theta K] T_p^{n+1} = [c - \Delta t(1-\theta)K]T^n - \Delta F$, in which C, K, and F are evaluated with temperatures determined from $T = \alpha T_p^{n+1} + (1-\alpha)T^n$, in which α is a constant between 0 and 1, and T_p^{n+1} is the corrected temperature. The value of θ in the predictor-corrector steps must be between 0 and 1, in which a value of 0 represents the forward difference method, a value of 1/2 represents the central difference method, a value of 2/3 represents the Galerkin method, and a value of 1 represents the backward difference method.

The forward difference method is an explicit method with first order accuracy in time. The forward difference method makes the system matrix diagonal and therefore quickly solvable, but the advantage gained in computing time with the diagonal matrix can be lessened or even eliminated by the stability requirement that the dimensionless time step be less than or equal to 1/2. The central difference scheme is the only two-level method with second order accuracy in time. Although the central difference method is unconditionally stable (for θ greater than or equal to 1/2, the discretization becomes unconditionally stable), the dimensionless time step must be small enough to prevent oscillation. The backward difference method is fully implicit and unconditionally stable, but is only first order accurate in time. The choice of method is left to the user and depends on the desired accuracy of the results.

All prevailing initial and boundary conditions must be included in the model and discretized accordingly. ProCAST has the ability to handle both Dirichlet and Neumann boundary conditions. Temperature and/or time varying thermophysical properties and heat transfer coefficients can also be included. Once all of the initial conditions, boundary conditions, thermophysical properties, and heat transfer coefficients are defined, ProCAST can use the method described above to predict the time-varying temperature profile of the casting and the mold.

4.3. Prior Application of Solidification Models

Several cases of the use of modeling to improve casting processes can be found in the literature. For example, modeling of mold filling, porosity formation, hot tearing, etc., has been successful in predicting casting structure and defect formation. Some problems for which modeling packages have been used successfully will be reviewed here.

Yu et al. modeled the investment casting of single-crystal nickel-based superalloy airfoils using PATRAN and IDEAS for geometry and mesh generation and graphical display and the finite element software TOPAZ/SDRC for solving the transient heat conduction problem.[19],[46] Their model was used to predict grain size, dendrite arm spacing, porosity formation, hot tearing, misruns, and grain defects. They successfully predicted that (1) the shroud area would have a high tendency for microporosity based on a low G:R ratio and (2) there would be no macroshrinkage. In addition, a high risk of hot tearing was predicted using a strain rate index that was related to casting contraction at various points. The casting process was modified based on these modeling results for hot crack tendency, and crack-free castings resulted.

The metal temperature distribution in the casting immediately after mold fill was also calculated to see if misruns were likely to occur in a certain casting. The model predicted correctly that misruns would not be a problem because the lowest temperature of the molten metal after filling was above the liquidus temperature of the alloy. Further modeling was conducted to predict risk areas for various grain defects specific to single crystal or directionally solidified castings. The model proved to be effective in predicting such defects.

Ohtsuka, Mizuno, and Yamada[47] simulated aluminum permanent mold casting (PMC) using a two dimensional explicit finite difference method to solve the transient heat conduction problem. Their goal was to determine the effect of different mold coatings and to use their results to improve the PMC process. In the original process for PMC of aluminum alloy wheels, shrinkage defects were found in the rim. The simulation confirmed that the rim section was cut off from the flow of molten metal before it solidified. Computer simulation was used to determine changes in mold coatings and adjustments in section size to eliminate the resulting shrinkage in the rim area. Upon implementing these changes, porosity-free castings were produced.

4.4. Input Data for Solidification Modeling

Solidification modeling cannot be successful without accurate input data. Many solidification models are derived from basic heat and fluid flow equations, which contain thermophysical properties and various other fitting parameters. If these parameters cannot be accurately determined, modeling attempts will be futile. Major effort has been exerted to determine experimentally the input data for modeling; some techniques used in such work are reviewed below. This review is divided into three separate sections according to the type of data being determined. The first section describes methods used to find thermophysical material properties such as density, thermal conductivity, and specific heat. Section two describes the interface heat transfer coefficient and its relation to modeling. Section three addresses other required input data for solidification modeling.

4.4.1. Thermophysical Properties

Solidification modeling often involves solving differential equations for heat transport such as the one-dimensional, unsteady heat conduction equation. The general differential equation for conductive heat transfer in one dimension is $\partial T/\partial t = \alpha(\partial^2 T/\partial x^2)$, in which $\alpha = k/\rho C$ is the thermal diffusivity, k is the thermal conductivity, ρ is the density, and C is the heat capacity of the material through which heat is being conducted. In order to solve this equation, the value of α must be known. Some problems require just α to be known, while others require the values of the components of α to be known separately. In many cases, α, k, ρ, and C must be known for both the mold and the metal being cast.

81

Because casting is not an isothermal process, any variation of α, k, ρ, and C with temperature must be known as well. Because phase transformations are involved, any change in properties due to such transformations must also be known. It is easy to see how many thermophysical data are required for solidification modeling. The accurate determination of such data is critical, and modelers should be aware of the source of their input data, the experimental method used, and the accuracy of the method used.

One method for determining thermal diffusivity is the laser flash diffusivity method.[48] A small disc-shaped sample of the material being tested is hit with a short laser burst on one face. The temperature rise on the opposite face due to the laser burst is measured. A computer collects the data and computes results, and then compares the data to a theoretical model. Solid density can be calculated from measured dimensions and mass. The accuracy of the density calculation is determined by the accuracy with which dimensions and mass are measured. A method for determining specific heat is differential scanning calorimetry.[48] This method can also be used to determine transformation energetics such as latent heat of fusion. Measuring liquid properties is more difficult due to the high temperatures and high reactivity involved; thus, data on liquid alloy properties are scarce.

4.4.2. Interface Heat Transfer Coefficients

The interface heat transfer coefficient (IHTC) is used to quantify the resistance of an interface to the transfer of heat. The IHTC h is defined by the equation $q = h(T_2 - T_1)$, in which h is the IHTC, T_1 and T_2 are the temperatures on either side of the interface, and q is the heat flux per unit area across the interface. The value of the IHTC depends on the nature of the interface. For cases in which the mating surfaces are in perfect atomic contact, the IHTC approaches infinity ($T_1 = T_2$) and the interface does not resist heat transfer. In this case, heat transfer across the interface is governed by the thermophysical properties of the materials making up the interface. For cases in which the two materials are not interfaced perfectly, a gap exists at the interface ($T_1 \neq T_2$). This gap considerably lowers the IHTC, and the interface resists heat transfer. The value of the IHTC decreases as the gap becomes larger or as an insulating oxide layer forms on the surface of the mating materials. As the IHTC decreases, the heat transfer across the interface becomes highly dependent on the interface characteristics and pressure and less dependent on the characteristics of the mating materials themselves.

82

In order to model the heat extraction that takes place in casting in metal molds (such as PMC, die casting, or casting against a chill), the metal-mold interface heat transfer coefficient must be known. The metal-mold IHTC is highly system dependent, and can vary with system configuration, mold temperature, metal temperature, applied pressure, mold surface finish, casting alloy, and mold material. As many of these properties change throughout the casting process, the IHTC also changes with time. Because the IHTC is affected by configuration, different sections of the interface can have different IHTCs as well.

In a typical permanent mold casting cycle, the value of the IHTC changes considerably. During mold filling, the IHTC is very high due to the flow of the liquid past the mold wall and the good contact between the liquid and the mold wall. When the molten alloy completely fills the mold, the liquid metal is in good contact with the mold and the IHTC remains fairly high. As soon as a solid skin forms, the contact between the casting and the mold deteriorates. The mold heats up and expands while the casting cools down and contracts. Depending on the mold configuration, this combined expansion and contraction can result in (1) the formation of a gap between the casting and the mold, which lowers the IHTC, or (2) an increase in the pressure forcing the mold and casting together, which raises the IHTC. If the casting operation is performed in air or with the use of an oxygen-containing lubricant, an oxide film can form on the surface of the casting or the mold, further reducing the IHTC. Finally, when the mold is opened, contact between the metal and mold ceases to exist, and the IHTC decreases again.

The value of the IHTC cannot be predicted from thermophysical properties of the casting alloy. Rather, the transient temperature distributions within the mold and the casting must be measured experimentally. Using these temperature data, (1) the value of the IHTC in a previously validated model is varied until the predicted temperature profiles match the measured temperature profiles, or (2) the temperature profile is input into the solidification model, which is then inverted to solve for the IHTC in a best-fit sense. Both analytical and numerical solutions to the solidification problem can be used in such approaches. Analytical solutions are only applicable in certain cases, and many simplifying assumptions are made as discussed previously in this chapter. Numerical methods can be used more broadly, and can be applied to nonlinear equations.

Many researchers have investigated heat transfer coefficients in solidification processes. A review of some methods and results from experiments involving heat transfer coefficients for permanent mold casting, die casting, and casting against a chill follows. A compilation of many of these results is contained in a report by Papai and Mobley, who also offer a simple model of heat transfer across interfaces.[49]

Sun studied the effect of casting alloy, mold material, and mold surface condition on the thermal resistance of casting-mold interfaces.[50] The experimental procedure consisted of plunging a rod of mold material into a molten bath of casting alloy (moving it around slowly to maintain contact with hot liquid) and recording the temperature history at the mid-radius of the rod with a thermocouple (Figure 53[50]). The mold materials were either cast iron, copper, graphite, molybdenum, or aluminum with various surface conditions such as "sand blasted," "graphite coated", "zircon coated", "polished and sand blasted", or "machine finished". The casting alloys used were Hastelloy X (bath temperature of 2600 °F) or pure aluminum (bath temperature of 1250 °F). The measured temperature histories were compared to calculated temperature histories. An explicit finite difference method was used to solve the equations for one dimensional heat flow in a cylindrical rod to obtain calculated temperature histories for various initial and boundary conditions for immersion times up to 30 seconds. Thermophysical properties of the mold materials were allowed to vary with temperature, and the IHTC was assumed to be constant or linearly dependent on time.

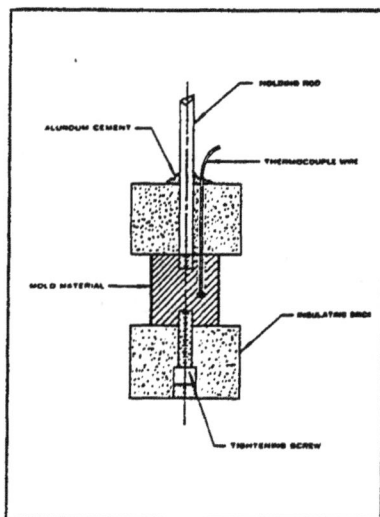

Figure 53. Arrangement of the sample rod.

Table 11. The simulated IHTC for various mold materials

Mold material	Surface condition	Molten bath temperature (°F)	Simulated IHTC* (Btu/ft²-hr-°F)	Simulated IHTC at various times (Btu/ft²-hr-°F)***			
				5 sec.	10 sec.	15 sec.	20 sec.
Cast iron	Sand blasted	1250	$h = 50 + 1.26 \times 10^5 t$	225	400	575	750
	Sand blasted	2600	$h = 50 + 1.26 \times 10^5 t$	225	400	575	750
	Graphite coated (1/32 ~ 1/16")	1250	$h = 50 + 1.8 \times 10^4 t$	75	100	125	175
	Graphite coated (very thin)	2600	$h = 50 + 9.0 \times 10^4 t$	175	300	425	550
	Zircon coated	2600	$h = 100 + 5.4 \times 10^4 t$	175	250	325	400
Copper	Polished & sand blasted	1250	$h = 100 + 9.0 \times 10^4 t$	225	350	475	600
	Polished & sand blasted	2600	$h = 100 + 9.0 \times 10^4 t$	225	350	475	600
	Graphite coated (3/64 ~ 1/16")	1250	$h = 40$ to 45**				
	Graphite coated (~1/32")	2600	$h = 100 + 7.2 \times 10^4 t$	200	300	400	500
	Zircon coated (~1/32") (~1/16") (~3/32")	2600	$h = 100 + 5.4 \times 10^4 t$ $h = 50 + 3.6 \times 10^4 t$ $h = 50 + 1.8 \times 10^4 t$	175 100 70	250 150 100	325 200 125	475 250 150
Graphite	Machine finished	1250 2600	$h = 100 + 2.7 \times 10^5 t$ $h = 100 + 2.7 \times 10^5 t$	475 475	850 850	1225 1225	1600 1600
Molybdenum	Zircon coated (~1/32") (~1/16")	1250	$h = 50 + 1.26 \times 10^4 t$ $h = 30 + 7.00 \times 10^3 t$	80 40	110 50	140 60	170 70
	Sand Blasted	2600	$h = 50 + 1.9 \times 10^5 t$	314	578	842	1106
	Zircon coated (~1/32") (~1/16")	2600	$h = 80 + 1.08 \times 10^4 t$ $h = 50 + 3.6 \times 10^3 t$	95 55	110 60	125 65	140 70
Aluminum	Sand blasted	1250	$h = 1.44 \times 10^5 t$	200	400	600	800
	Graphite coated	1250	$h = 10 + 1.26 \times 10^4 t$	37.5	45	62.5	80

* t is the metal and mold surface contact time in hours, and is valid before the gap is formed, say <30 seconds (1/120 hr). ** May have bad thermocouple wire contact. *** To convert from Btu/hr-ft²-°F to kW/m²-K, multiply by 5.678 x 10⁻³.

The equations for the IHTC that provide the best fit to the experimental data for each combination of molten alloy, rod material, and rod surface condition are shown in Table 11.[50] Values of the IHTC range from 0.16 to 2.70 kW/m²-K after five seconds and from 0.40 to 9.08 kW/m²-K after twenty seconds. The results also suggest that (1) interface contact pressure plays a large role in determining the IHTC, (2) mold surface temperature and molten alloy temperature do not affect the IHTC in the early stages of casting, (3) coatings decrease the IHTC with thicker coatings resulting in lower values of the IHTC, and (4) conduction is the dominant mode of heat transfer in the early stages of casting. However, it must be realized that these various trends are very specific to the test method which was utilized. For example, the rod expanded as it was heated, and the solidifying alloy contracted as it cooled; thus the contact pressure at the interface between the two increased; this is not the typical case in an actual casting. Furthermore, because the contact pressure increased with time in Sun's experiments, no air gap was formed at the interface, and, consequently, the IHTC increased with time. In real cases in which an air gap does form, the IHTC decreases with time and radiative and convective heat flow do affect the IHTC, and, consequently, mold and metal temperatures affect the IHTC.

Nishida and Matsubara[51] studied the effect of pressure on the IHTC for casting aluminum in a cylindrical carbon steel mold. The experimental procedure comprised measurement of the temperature history at four places in the mold and three places in the casting during the casting cycle while applying a set load with either a universal tester or a squeeze casting machine (Figure 54[51]). A numerical model using constant thermophysical properties was used to predict the values of the IHTC from the measured temperature profiles. The results are shown in Figures 55 and 56.[51] The minimum IHTC varied from a value less than 4 kW/m^2-K with no applied pressure to a value of more than 40 kW/m^2-K with ~100 MPa applied pressure (Figure 57[51]).

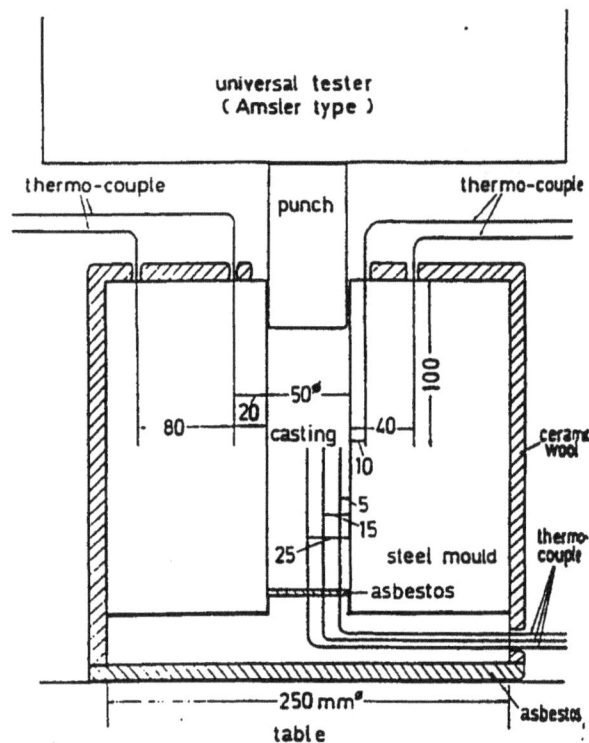

Figure 54. Apparatus for casting and measurement of temperature.

86

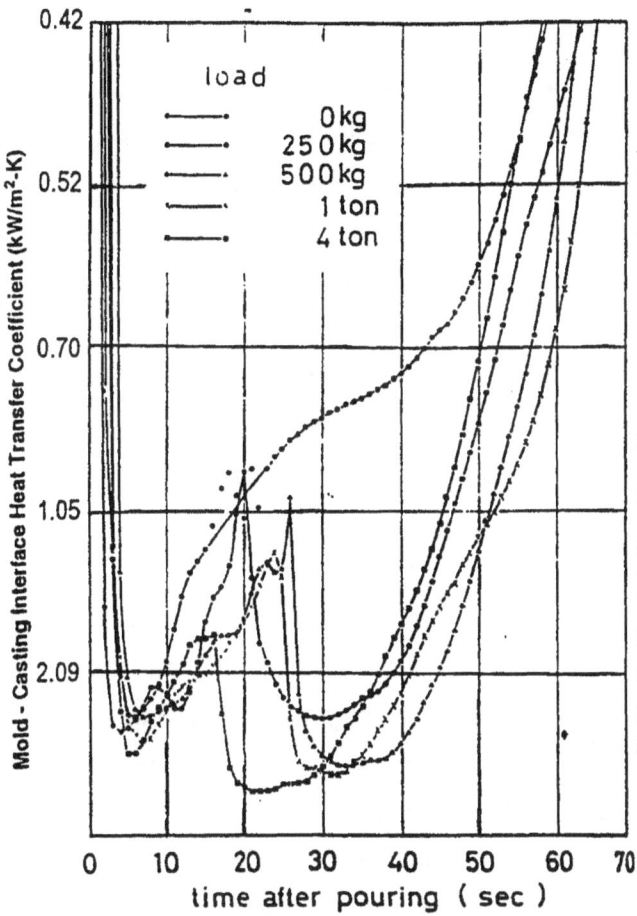

Figure 55. Thermal resistance curves versus time after pouring under various thermal loads. (Steep points indicate start of load.)

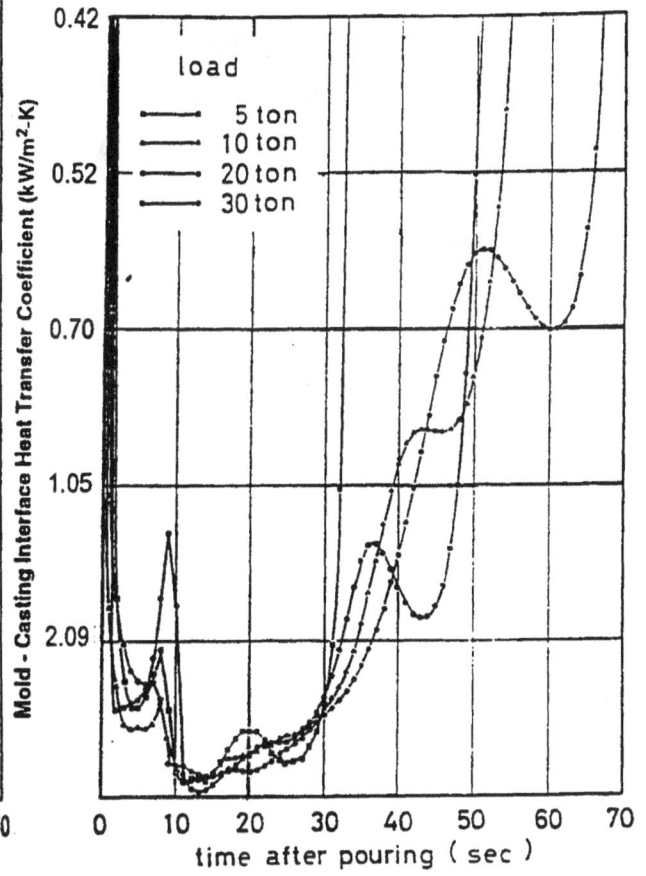

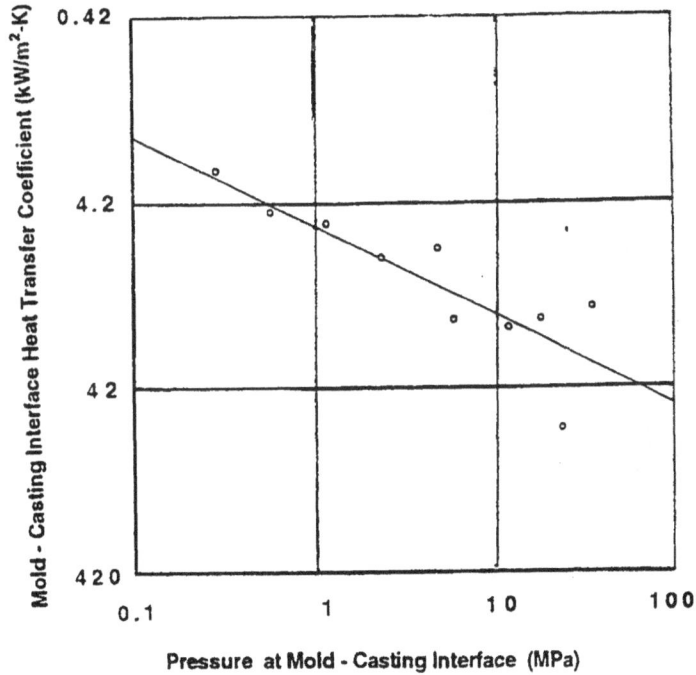

Figure 56. Thermal resistance curves versus time after pouring under various thermal loads. (Steep points indicate start of load.)

Figure 57. Minimum values of thermal resistance versus pressure.

87

Nishida, Droste, and Engler[52] studied the effect of shape on the IHTC for pure aluminum and Al-13.2%Si cast in graphite coated molds of two different geometries - a cylindrical mold and a flat mold (Figures 58 and 59[52]). Their main objective was to study the effect of air gap formation at the mold-metal interface on the IHTC. They used quartz coupling rods to measure the displacement of the casting and the mold separately. They also measured the temperature history at several points within the mold and the casting using thermocouples. For the cylindrical castings, their results revealed that the mold expanded as it heated up, and no contraction was seen during the casting cycle (Figures 60 and 61[52]). The cylindrical castings did not seem to contract and pull away from the mold until solidification was nearly complete. The corresponding heat transfer coefficients reached a maximum of ~3 kW/m^2-K after ~10 seconds and did not begin to decrease sharply until the casting began to contract significantly. By contrast, for the flat castings, the mold seemed to contract just after pouring and then gradually expanded back toward its original size (Figures 62 and 63[52]). The casting contracted with the mold as well and thus an air gap did not form until the mold began to expand. Correspondingly, the heat transfer coefficient increased to a maximum until the air gap formed and then decreased sharply. The value of the maximum heat transfer coefficient for the flat casting depended on the level of mold constraint. In a weakly constrained mold the maximum was found to be ~4 kW/m^2-K, while in a tightly constrained mold the maximum was found to be ~3 kW/m^2-K. The data for both mold geometries also demonstrated that alloy composition plays only a small role in determining the heat transfer coefficient.

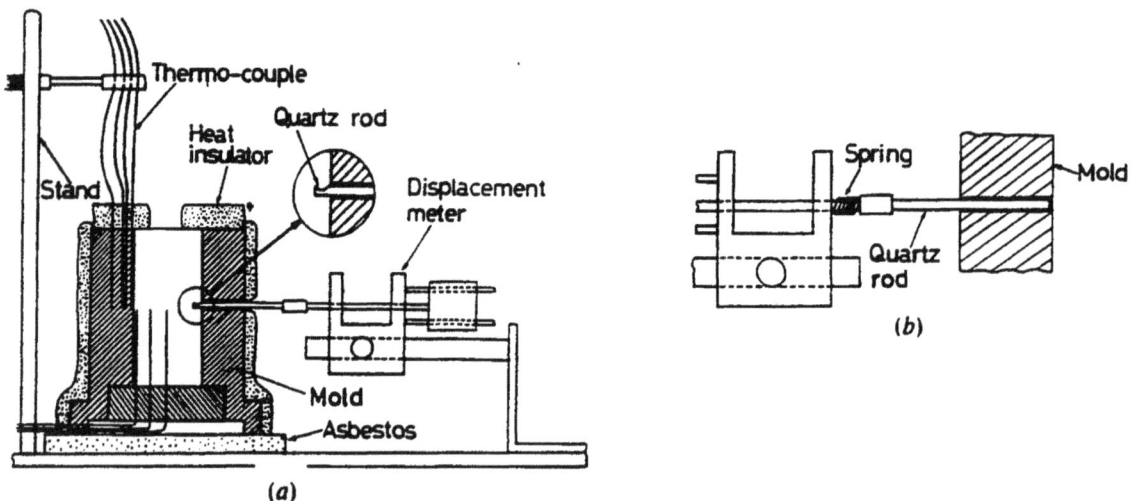

Figure 58. (a) Apparatus for measurement of casting displacement and temperature; (b) Quartz coupling rod and mold for measurement of mold displacement.

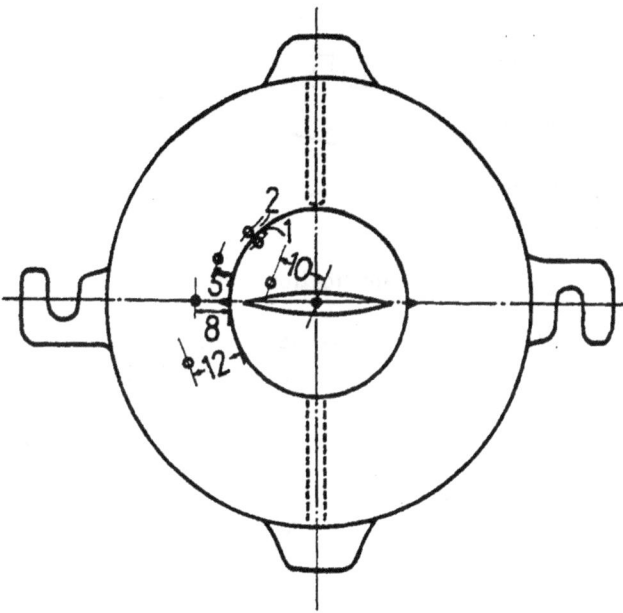

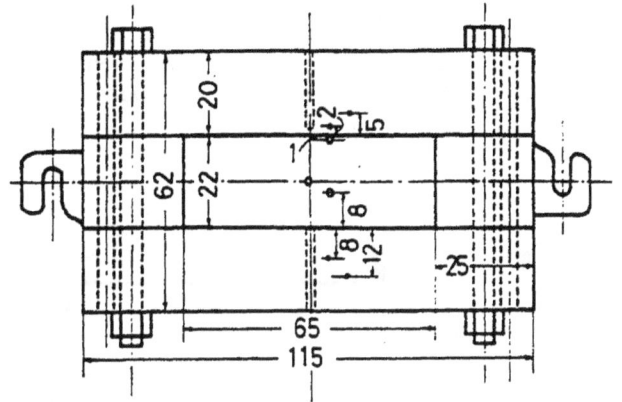

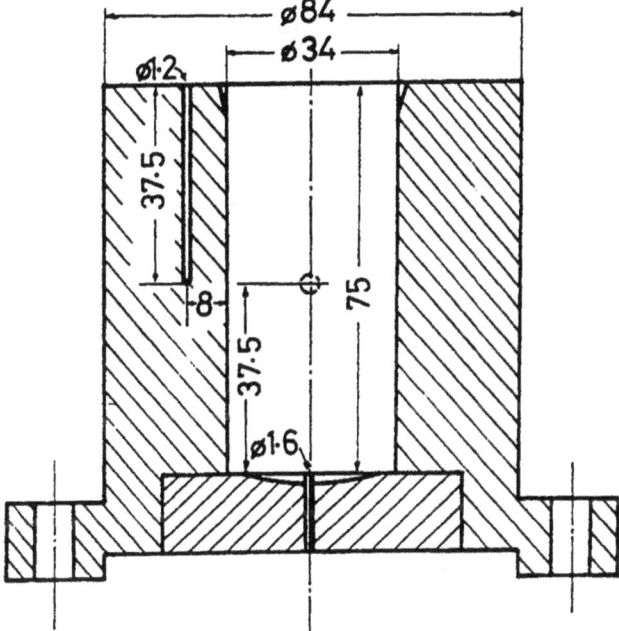

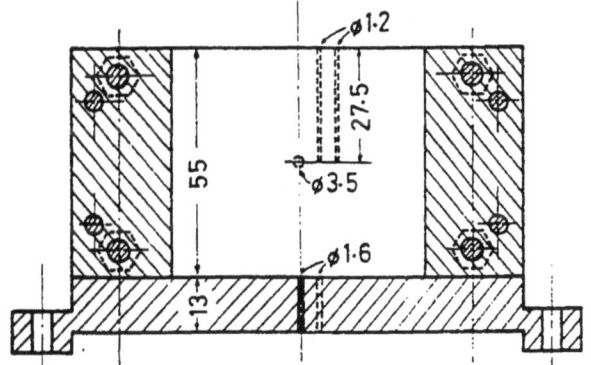

Figure 59. Molds used for the measurement of displacements and temperatures: (a) mold for cylindrical casting, and (b) mold for flat casting.

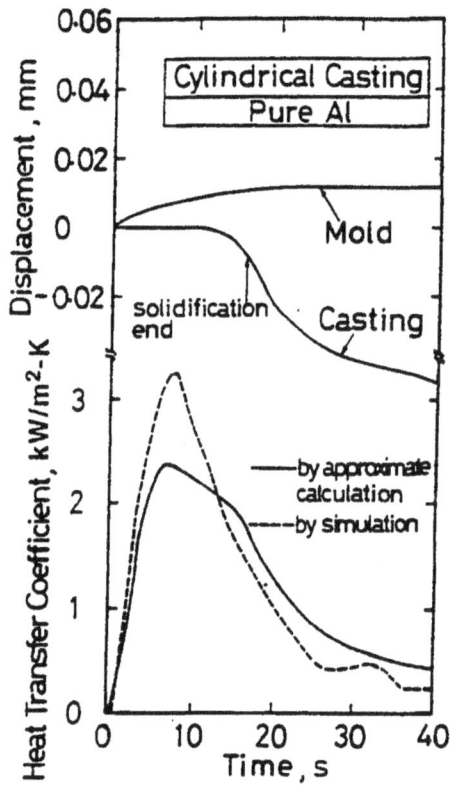

Figure 60. Heat transfer coefficient (lower curves) compared with mold and casting displacements (upper curves) for cylindrical pure aluminum casting.

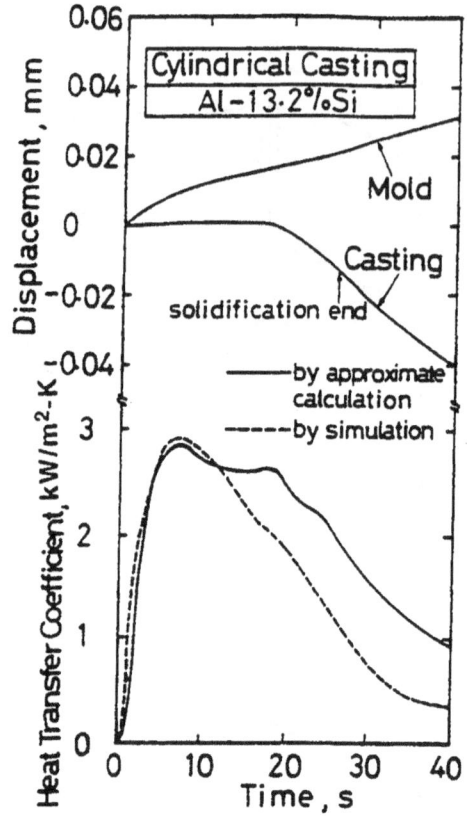

Figure 61. Heat transfer coefficient (lower curves) compared with mold and casting displacements (upper curves) for cylindrical Al-13.2%Si alloy casting.

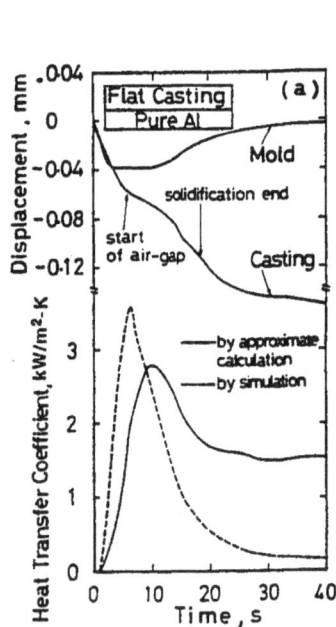

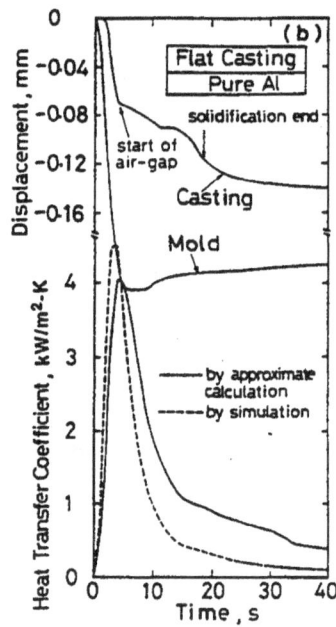

Figure 62. Heat transfer coefficient (lower curves) compared with mold and casting displacements (upper curves) for flat pure aluminum casting; (a) for a tightly constrained mold, and (b) for a weakly constrained mold.

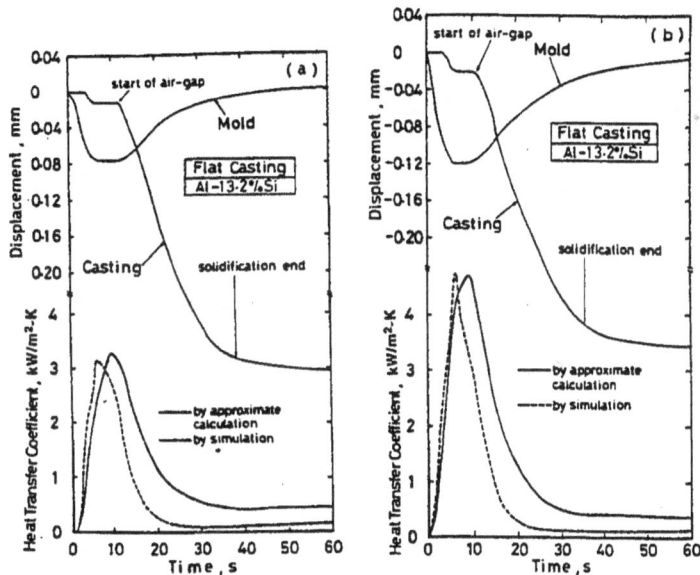

Figure 63. Heat transfer coefficient (lower curves) compared with mold and casting displacements (upper curves) for flat Al-13.2%Si alloy casting; (a) for a tightly constrained mold, and (b) for a weakly constrained mold.

Sully studied the effect of casting size, casting alloy, mold material, and mold geometry on the IHTC.[53] Plate castings of various materials and sizes were cast in thermocoupled horizontal sand molds with bottom plate chills, vertical permanent molds, and sand molds with pipe chills (Figures 63 - 66[53]). Both trial-and-error (using an implicit finite difference method) and inverse heat transfer methods were used to determine IHTCs for the various casting configurations. The results were used to draw conclusions about the effect of casting size, casting geometry, mold material, cast alloy, and casting and mold surface temperatures on the IHTC.

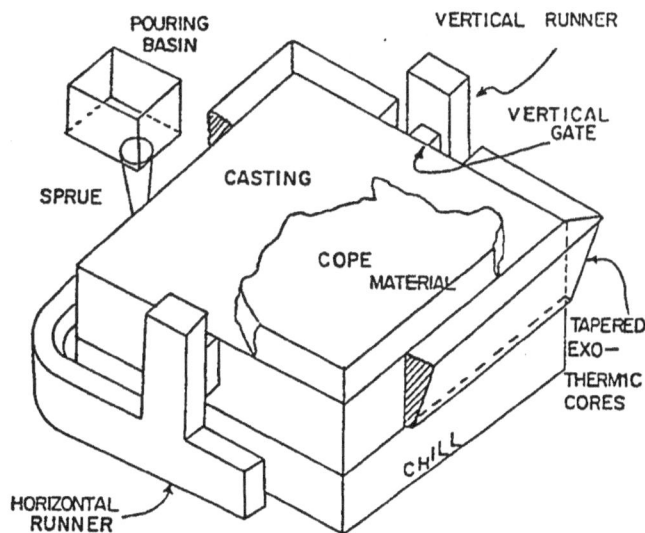

Figure 64. Mold arrangement to produce unidirectionally solidified 20" x 20" x 5.5" steel plates.

91

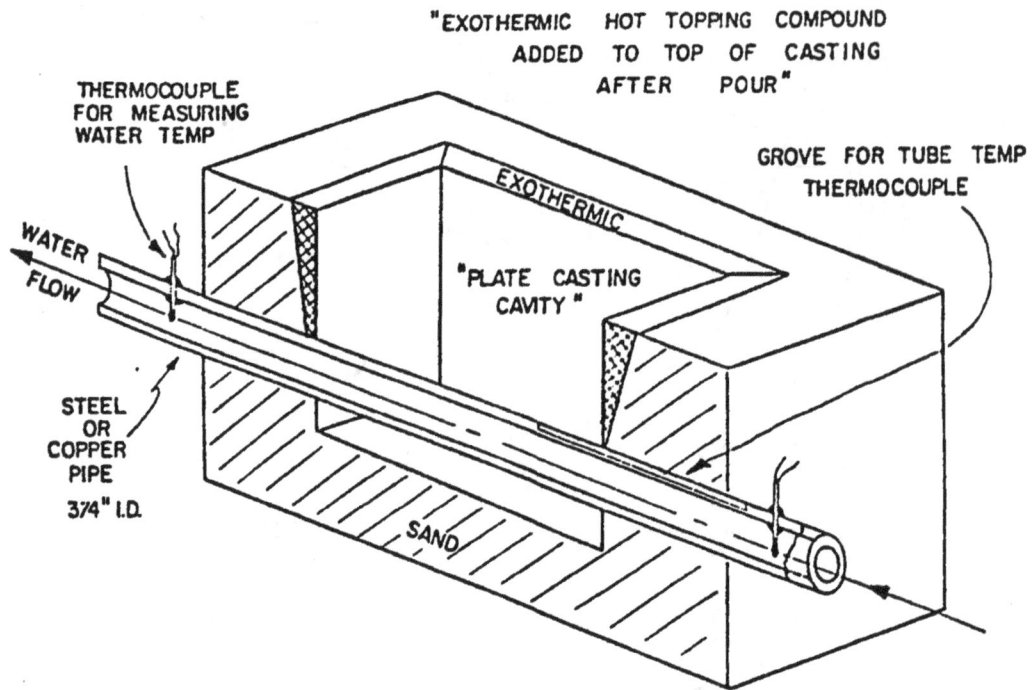

THERMOCOUPLE
FOR MEASURING
WATER TEMP

"EXOTHERMIC HOT TOPPING COMPOUND
ADDED TO TOP OF CASTING
AFTER POUR"

GROVE FOR TUBE TEMP
THERMOCOUPLE

WATER
FLOW

EXOTHERMIC

"PLATE CASTING
CAVITY"

STEEL
OR
COPPER
PIPE
3/4" I.D.

SAND

Figure 65. Cross-section of the 10" x 10" x 5.5" mold for small experiments with water-cooled pipe chills.

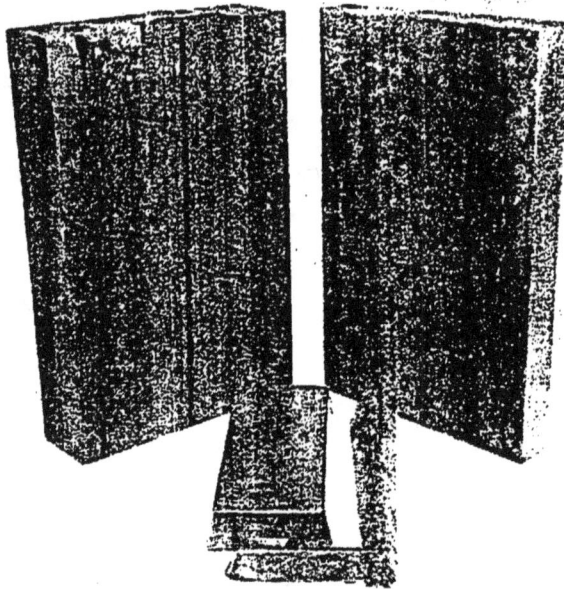

Figure 66. Mold used to vertically pour 6" x 2" x 0.75" plate casting. (The flat side of the mold is of variable material and is instrumented with thermocouples.)

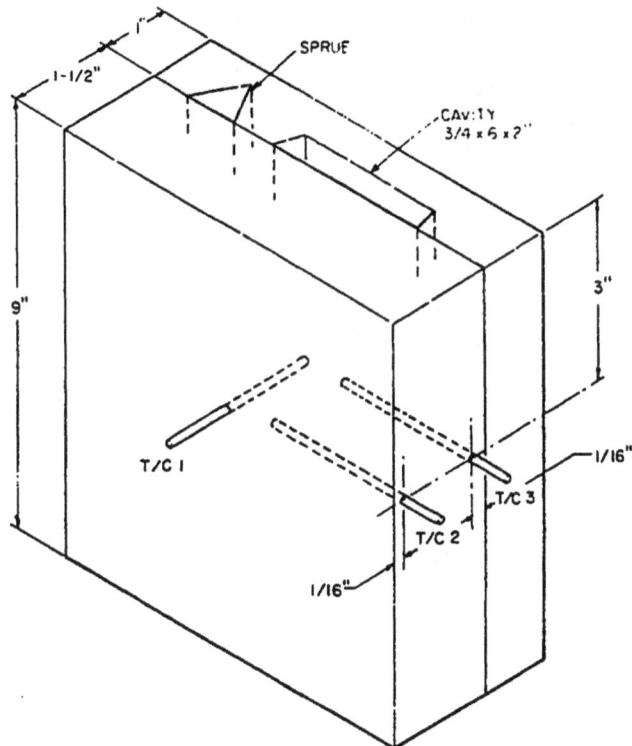

Figure 67. Location of three 28-gage chromel-alumel thermocouples on the flat half of the vertical permanent mold.

The IHTCs determined by Sully are shown in Figures 68 to 73[53] for the various casting geometries. All of the plate castings without a pipe chill exhibited similar variations of the IHTC with time. A peak in the IHTC was reached during or slightly after completion of mold fill after which the IHTC declined rapidly and eventually reached a steady state value. The height, width, and shape of the peak and the eventual steady state the IHTC value varied from one configuration to another. The peak values ranged from 0.68 to 1.02 kW/m^2-K and the steady state values range from 0.11 to 0.57 kW/m^2-K.

General conclusions from Sully's research are as follows: (1) geometry affects the IHTC significantly, and, by comparison, the mold material and the casting alloy have only a small effect on the IHTC, (2) casting size controls the temporal variation of the IHTC, and (3) casting surface temperature has a large effect on the IHTC, while the mold surface temperature does not.

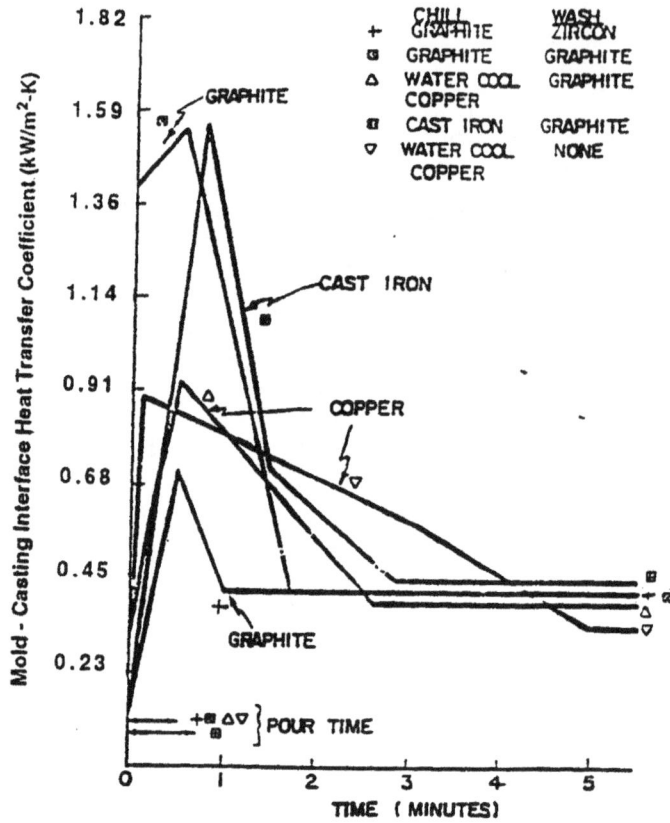

Figure 68. Casting-chill interface heat transfer coefficients for five plate chill-chill wash combinations.

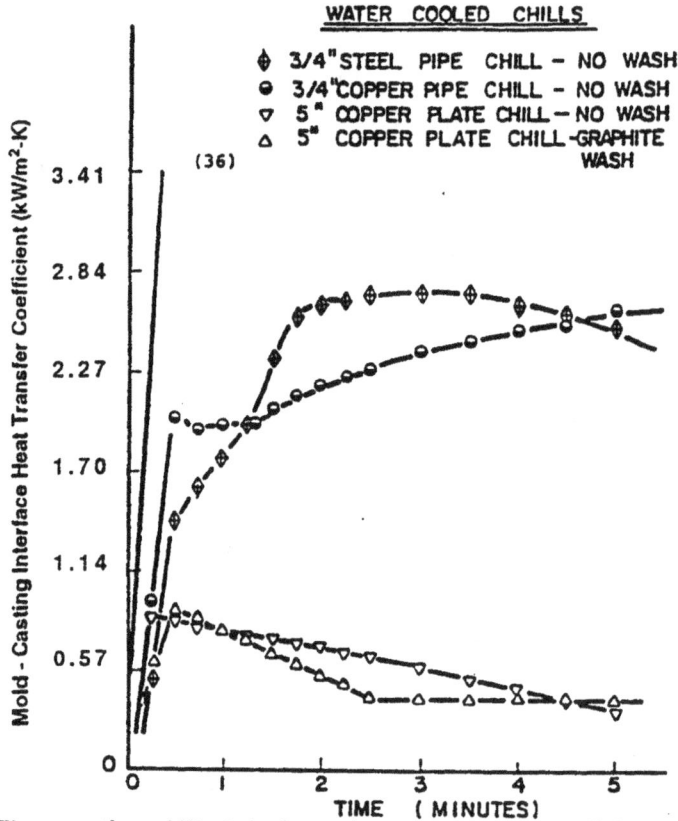

Figure 69. The casting-chill interface heat transfer coefficient as a function of time for water cooled pipe chills and water-cooled plate chills.

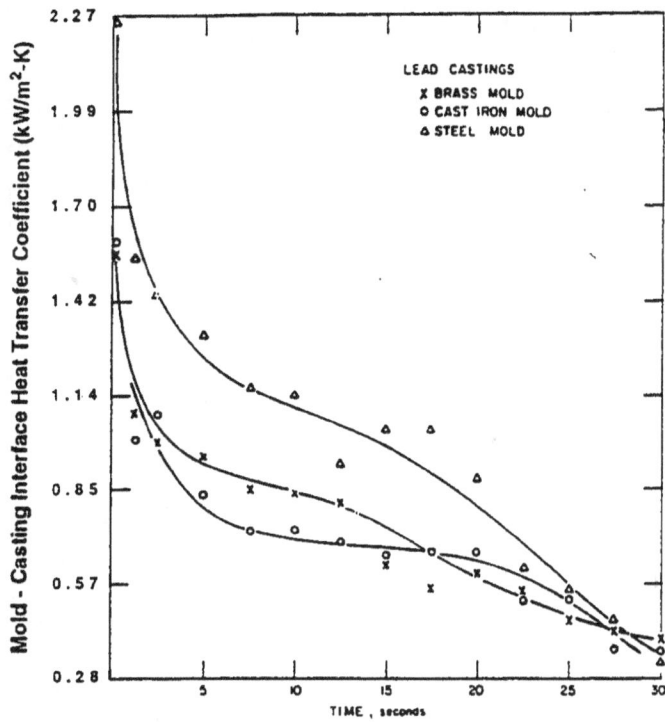

Figure 70. The heat transfer coefficient between lead castings and vertical permanent molds.

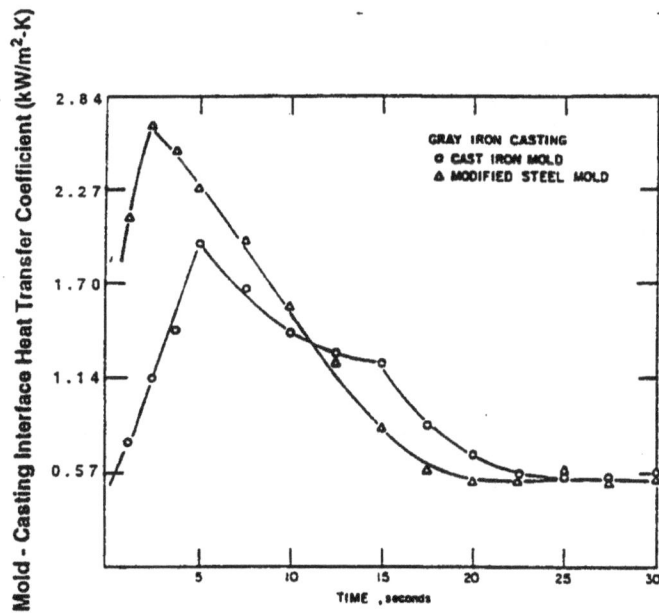

Figure 71. Heat transfer coefficient between a 2" x 6" x 0.75" gray cast iron plate in a vertical permanent mold.

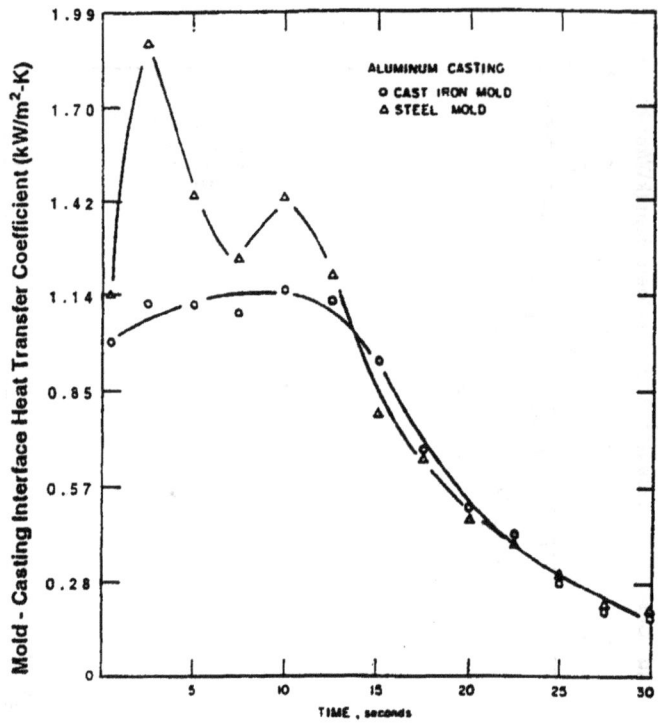

Figure 72. The heat transfer coefficient between aluminum castings and vertical permanent molds.

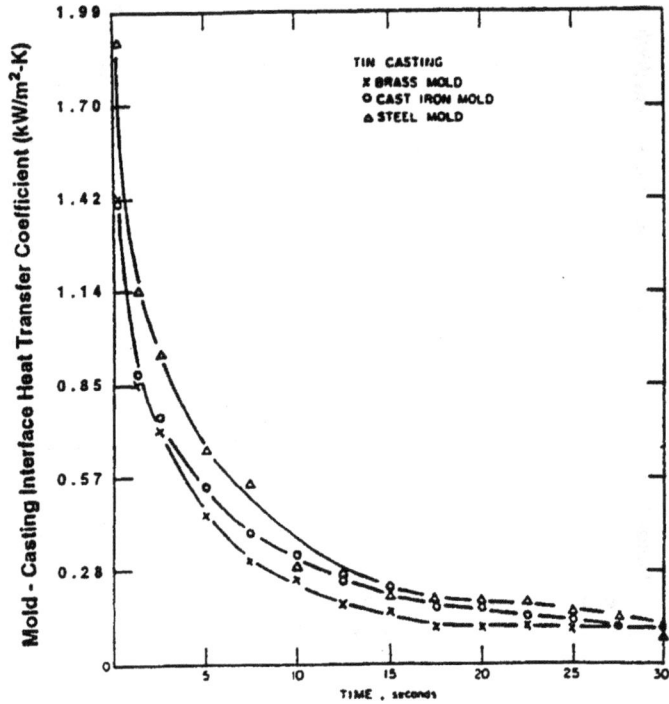

Figure 73. The heat transfer coefficient between tin castings and vertical permanent molds.

4.4.3. Other Input Parameters for Modeling of Casting

Other parameters that are sometimes required for modeling include viscosity, emissivity, mass diffusivity, and creep/plastic properties. Viscosity is required for fluid flow modeling, emissivity is required for radiation heat transfer modeling, mass diffusivity is required for mass transfer modeling, and creep/plastic properties are required for deformation and fracture modeling. Some of these properties may be found in the literature, but many are temperature and/or microstructure dependent and thus must often be measured for advanced casting models.

4.5. Discussion of Research Issues/Approaches

Research issues in modeling of solidification in PMC of titanium alloys include:
- Determination of heat transfer coefficients for specific systems
- Determination of thermophysical properties for specific alloys
- Validation of numerical models with actual castings
- Use of a validated model to solve real problems

Work needs to be done to achieve the capability of accurately modeling permanent mold casting of titanium alloys. ProCAST software could be used to model several castings, and casting trials could be conducted in thermocoupled molds to determine mold-metal heat transfer coefficients and to validate the modeling results. Results from the microstructure modeling research could be included in ProCAST or similar software to gain the capability of automatically predicting microstructural features based on modeling results.

97

5. References

1. J. D. Destefani, "Introduction to Titanium and Titanium Alloys," <u>Metals Handbook</u>, Tenth Edition, v. 2, p. 588, 1990.

2. Lampman, S., "Wrought Titanium and Titanium Alloys," <u>Metals Handbook</u>, Tenth Edition, v. 2, pp. 601-623, 1990.

3. Donachie Jr., M. J., ed., <u>Titanium: A Technical Guide</u>, ASM International, 1988.

4. Eylon, D., J. R. Newman, and J. K. Thorne, "Titanium and Titanium Alloy Castings," <u>Metals Handbook</u>, Tenth Edition, v. 2, pp. 637-644, 1990.

5. Ford, D. A., "Casting Technology," <u>The Development of Gas Turbine Materials</u>, G. W. Meetham, ed., Halsted Press, NY, p. 169, 1981.

6. Colvin, G. N., "Titanium Alloys: Cast," <u>The Encyclopedia of Advanced Materials</u>, D. Bloor, ed., Oxford Pergamon, p. 2869, Nov. 1994.

7. Horton, R. A., "Investment Casting, " <u>Metals Handbook</u>, Ninth Edition, v. 15, pp. 253-269, 1988.

8. Froes, F. H., D. Eylon, and H. B. Bomberger, "Melting, Casting, and Powder Metallurgy," A Lesson from <u>Titanium and its Alloys</u>, ASM-MEI Course 27, Lesson 8, ASM International, Metals Park, OH, p. 13, 1994.

9. Eridon, J. M., "Hot Isostatic Pressing of Castings," <u>Metals Handbook</u>, Ninth Edition, v. 15, p. 538, 1988.

10. Colvin, G. N., D. W. Anderson, and R. J. Schmalholz, "Permanent Mold Casting of Titanium Aerospace Components."

11. West, C. E., and T. E. Grubach, "Permanent Mold Casting," <u>Metals Handbook</u>, Ninth Edition, v. 15, pp. 275-285, 1988.

12. Shivpuri, R., and S. L. Semiatin, "Wear of Dies and Molds in Net Shape Manufacturing," <u>Engineering Research Center for Net Shape Manufacturing</u>, Report No. ERC/NSM-88-05, June 1988.

13. Jones, P. E., W. J. Porter III, D. Eylon, and G. Colvin, "Development of A Low Cost Permanent Mold Process for TiAl Automotive Valves," <u>Proceedings of ISTGA</u>, 1995.

14. Anderson, D., unpublished research, Mar. 1994.

15. Anderson, D., unpublished research, Jan. 1995.

16. Giamei, A. F., "Solidification Process Modeling: Status and Barriers," <u>JOM</u>, v. 45 (1), p.51.

17. Tsumagari, N., C. E. Mobley, and P. R. Gangasani, "Construction and Application of Solidification Maps for A356 and D357 Aluminum Alloys," <u>AFS Transactions</u>, v. 119, pp. 335-341, 1993.

18. Overfelt, T., "The Manufacturing Significance of Solidification Modeling," <u>JOM</u>, v. 44 (6), pp. 17-20, 1992.

19. Yu, K. O., M. J. Beffel, M. Robinson, D. D. Goettsch, B. G. Thomas, D. Pinella, and R. G. Carlson, "Solidification Modeling of Single-Crystal Investment Castings," <u>AFS Transactions</u>, v. 53, pp. 417-428, 1990.

20. Tu, J. S., and R. K. Foran, "The Application of Defect Maps in the Process Modeling of Single-Crystal Investment Casting," <u>JOM</u> v. 44 (6), pp. 26-28, 1992.

21. Purvis, A. L., C. R. Hanslits, and R. S. Diehm, "Modeling Characteristics for Solidification in Single-Crystal Investment-Cast Superalloys," <u>JOM</u> v. 46 (1), pp. 3 8 - 41, 1994.

22. Hunt, J. D., "Steady State Columnar and Equiaxed Growth of Dendrites and Eutectic," <u>Materials Science and Engineering</u>, v. 65, pp. 73-83, 1984.

23. Kurz, W., and D. J. Fisher, <u>Fundamentals of Solidification</u>, Trans Tech Publications, 1984.

24. Spear, R. E., and G. R. Gardner, "Dendrite Cell Size," <u>AFS Transactions</u>, v. 71, p. 209, 1963.

25. Fang, Q., and D. Granger, "Porosity Formation in Modified and Unmodified A356 Alloy Castings," <u>AFS Transactions</u>, v. 97, p. 42, 1983.

26. Elliot, R., <u>Eutectic Solidification Processing</u>, Butterworths & Co., 1983.

27. Jackson, K. A., and J. D. Hunt, "Lamella and Rod Eutectic Growth," <u>Transactions of AIME</u>, v. 236, p. 1129, 1966.

28. Trivedi, R., and W. Kurz, "Microstructure Selection in Eutectic Alloy Systems," <u>Solidification Processing of Eutectic Alloys</u>, Stefanescu et al., eds., TMS, 1988.

29. McLean, M., <u>Directionally Solidified Materials for High Temperature Service</u>, London: TMS, p. 26, 1983.

30. Bouse, G. K., and J. R. Mihalisin, "Metallurgy of Investment Cast Superalloy Components," <u>Superalloys, Supercomposites, and Superceramics</u>, J. K. Tien and T. Caulfield, eds., Boston, MA: Academic Press, pp. 149-182, 1989.

31. Niyama, E., et al., " A Method of Shrinkage Prediction and Its Application to Steel Casting Practice," <u>AFS International Cast Metals Journal</u>, p. 52, Sept. 1982.

32. Lecompte-Beckers, J., "Study of Microporosity formation in Nickel-Base Superalloys," <u>Metallurgical Transactions A.</u>, v. 19A, pp. 2341-2348, Sept. 1988.

33. Flemings, M. C., <u>Solidification Processing</u>, McGraw-Hill, Inc., 1974.

34. McCullough, C., J. J. Valencia, C. G. Levi, and R. Mehrabian, "Phase Equilibria and Solidification in Ti-Al Alloys," <u>Acta Metallurgica</u>, v. 37 (5), pp. 1321-1336, 1989.

35. Nurminen, J. I., and H. D. Brody, "Dendrite Morphology in Titanium Base Alloys, Science and Technology of Titanium, pp. 1893-1914, 1972.

36. Soffa, L. L., "Titanium Castings for Airframe Structural Applications," Metals Engineering Quarterly, p. 33, August 1968.

37. Froes, F. H., and R. G. Rowe, "Rapidly Solidified Titanium - A Review," Titanium Rapid Solidification Technology, F. H. Froes and D. Eylon, eds., Warrendale, PA, pp. 1 - 13, 1986.

38. Moll, J. E., and C. F. Yolton, "Production and Characterization of Rapidly Solidified Titanium and Other Alloy Powders Made by Gas Atomization," Titanium Rapid Solidification Technology, F. H. Froes and D. Eylon, eds., Warrendale, PA: TMS, pp. 5 2 - 53, 1986.

39. Eylon, D., and F. H. Froes, "HIP Compaction of Titanium Alloy Powders at High Pressure and Low Temperature (HPLT)," Titanium Rapid Solidification Technology, F. H. Froes and D. Eylon, eds., Warrendale, PA: TMS, p. 274, 1986.

40. Vogt, R. G., D. Eylon, and F. H. Froes, "Effect of Er, Si, and W Additions on Powder Metallurgy High Temperature Titanium Alloys," Titanium Rapid Solidification Technology, F. H. Froes and D. Eylon, eds., Warrendale, PA: TMS, p. 180, 1986.

41. Yolton, C. F., T. Lizzi, V. K. Chandhok, and J. H. Moll, "Powder Metallurgy of Titanium Aluminide Components," Titanium Rapid Solidification Technology, F. H. Froes and D. Eylon, eds., Warrendale, PA: TMS, p. 265, 1986.

42. Bomberger, H. B., and F. H. Froes, "Prospects for Developing Novel Titanium Alloys Using Rapid Solidification," Titanium Rapid Solidification Technology, F. H. Froes and D. Eylon, eds., Warrendale, PA: TMS, pp. 21-39, 1986.

43. Gaskell, D. R., An Introduction to Transport Phenomena in Materials Engineering, Macmillan Publishing Company, New York, NY, pp. 401 - 416, 1992.

44. Carslaw, H. S., and J. C. Jaeger, Conduction of Heat in Solids, 2nd ed., Oxford University Press, London, 1986.

45. ProCAST User's Manual, Appendix A, "Mathematical Formulations," pp. 282-308.

46. Yu, K. O., J. J. Nichols, and M. Robinson, "Finite-Element Thermal Modeling of Casting Microstructures and Defects," JOM, v. 44 (6), pp. 21-25, 1992.

47. Ohtsuka, Y., K. Mizuno, J. Yamada, "Application of a Computer Simulation System to Aluminum Permanent Mold Castings," AFS Transactions, v. 90, pp. 635-646, 1982.

48. Henderson, J. B., and H. Groot, "Thermophysical Properties of Titanium Alloys," A Report to UES, Inc., TPRL 1284, May 1993.

49. Papai, J., and C. Mobley, "Heat Transfer Coefficients for Solidifying Systems," Engineering Research Center for Net Shape Manufacturing, Report No. ERC/NSM-887-13, August 1987.

50. Sun, R. C., "Simulation and Study of Surface Conductance for Heat Flow in the Early Stage of Casting," AFS Cast Metal Research Journal, pp. 105-110, Sept. 1970.

51. Nishida, Y., H. Matsubara, "Effect of Pressure on Heat Transfer at the Metal Mould-Casting Interface," British Foundryman, v. 69, pp. 274-278, 1976.

52. Nishida, Y., W. Droste, and S. Engler, "The Air Gap Formation Process at the Casting-Mold Interface and the Heat Transfer Mechanisms Through the Gap," Metallurgical Transactions B, v. 17B, pp. 83 -844, 1986.

53. Sully, L. J. D., "The Thermal Interface Between Castings and Chill Molds," AFS Transactions, v. 84, pp. 735-744, 1976.